Maritime Security Partnerships

Committee on the "1,000-Ship Navy"—
A Distributed and Global Maritime Network

Naval Studies Board

Division on Engineering and Physical Sciences

NATIONAL RESEARCH COUNCIL
OF THE NATIONAL ACADEMIES

THE NATIONAL ACADEMIES PRESS
Washington, D.C.
www.nap.edu

THE NATIONAL ACADEMIES PRESS 500 Fifth Street, N.W. Washington, DC 20001

NOTICE: The project that is the subject of this report was approved by the Governing Board of the National Research Council, whose members are drawn from the councils of the National Academy of Sciences, the National Academy of Engineering, and the Institute of Medicine. The members of the committee responsible for the report were chosen for their special competences and with regard for appropriate balance.

This study was supported by Contract No. N00014-05-G-0288, DO 15 between the National Academy of Sciences and the Department of the Navy. Any opinions, findings, conclusions, or recommendations expressed in this publication are those of the author(s) and do not necessarily reflect the views of the organizations or agencies that provided support for the project.

International Standard Book Number-13: 978-0-309-11261-1
International Standard Book Number-10: 0-309-11261-3

Copies of this report are available from:

Naval Studies Board, National Research Council, The Keck Center of the National Academies, 500 Fifth Street, N.W., Room WS904, Washington, DC 20001; and

The National Academies Press, 500 Fifth Street, N.W., Lockbox 285, Washington, DC 20055; (800) 624-6242 or (202) 334-3313 (in the Washington metropolitan area); Internet, http://www.nap.edu.

Printed in the United States of America

THE NATIONAL ACADEMIES

Advisers to the Nation on Science, Engineering, and Medicine

The **National Academy of Sciences** is a private, nonprofit, self-perpetuating society of distinguished scholars engaged in scientific and engineering research, dedicated to the furtherance of science and technology and to their use for the general welfare. Upon the authority of the charter granted to it by the Congress in 1863, the Academy has a mandate that requires it to advise the federal government on scientific and technical matters. Dr. Ralph J. Cicerone is president of the National Academy of Sciences.

The **National Academy of Engineering** was established in 1964, under the charter of the National Academy of Sciences, as a parallel organization of outstanding engineers. It is autonomous in its administration and in the selection of its members, sharing with the National Academy of Sciences the responsibility for advising the federal government. The National Academy of Engineering also sponsors engineering programs aimed at meeting national needs, encourages education and research, and recognizes the superior achievements of engineers. Dr. Charles M. Vest is president of the National Academy of Engineering.

The **Institute of Medicine** was established in 1970 by the National Academy of Sciences to secure the services of eminent members of appropriate professions in the examination of policy matters pertaining to the health of the public. The Institute acts under the responsibility given to the National Academy of Sciences by its congressional charter to be an adviser to the federal government and, upon its own initiative, to identify issues of medical care, research, and education. Dr. Harvey V. Fineberg is president of the Institute of Medicine.

The **National Research Council** was organized by the National Academy of Sciences in 1916 to associate the broad community of science and technology with the Academy's purposes of furthering knowledge and advising the federal government. Functioning in accordance with general policies determined by the Academy, the Council has become the principal operating agency of both the National Academy of Sciences and the National Academy of Engineering in providing services to the government, the public, and the scientific and engineering communities. The Council is administered jointly by both Academies and the Institute of Medicine. Dr. Ralph J. Cicerone and Dr. Charles M. Vest are chair and vice chair, respectively, of the National Research Council.

www.national-academies.org

COMMITTEE ON THE "1,000-SHIP NAVY"— A DISTRIBUTED AND GLOBAL MARITIME NETWORK

ROBERT B. PIRIE JR., Chevy Chase, Maryland, *Co-chair*
DAVID A. WHELAN, The Boeing Company, *Co-chair*
NOEL K. CUNNINGHAM, Glendora, California
HENRY H. GAFFNEY, The CNA Corporation
GUNTHER HANDL, Tulane University Law School
JOHN T. HANLEY,[1] Institute for Defense Analyses
THOM J. HODGSON, North Carolina State University
JAMES D. HULL, Annapolis, Maryland
HARRY W. JENKINS JR., Gainesville, Virginia
CATHERINE M. KELLEHER, University of Maryland and Brown University
JERRY A. KRILL, Applied Physics Laboratory, Johns Hopkins University
THOMAS V. McNAMARA, Textron Systems
HEIDI C. PERRY, Charles Stark Draper Laboratory, Inc.
GENE H. PORTER, Nashua, New Hampshire
JOHN S. QUILTY, Oakton, Virginia
J. PAUL REASON, Washington, D.C.
NILS R. SANDELL JR., BAE Systems
H. EUGENE STANLEY, Boston University
JOHN P. STENBIT, Oakton, Virginia
ELIHU ZIMET, Gaithersburg, Maryland

Staff

CHARLES F. DRAPER, Director
ARUL MOZHI, Study Director (as of May 19, 2007)
EUGENE J. CHOI, Study Director (through May 18, 2007)
RAYMOND S. WIDMAYER, Senior Program Officer
IAN M. CAMERON, Associate Program Officer (through May 21, 2007)
SUSAN G. CAMPBELL, Administrative Coordinator
MARY G. GORDON, Information Officer
SEKOU O. JACKSON, Senior Project Assistant
SEYMOUR J. DEITCHMAN, Consultant
SIDNEY G. REED JR., Consultant

[1]John T. Hanley served on the committee from November 28, 2006, to August 5, 2007.

NAVAL STUDIES BOARD

Marine Corps Liaison Representative

LTGEN JAMES F. AMOS, USMC, Commanding General, Marine Corps Combat Development Command (through July 2, 2008)
LTGEN GEORGE J. FLYNN, USMC, Commanding General, Marine Corps Combat Development Command (as of July 28, 2008)

Staff

CHARLES F. DRAPER, Director
ARUL MOZHI, Senior Program Officer
RAYMOND S. WIDMAYER, Senior Program Officer
BILLY M. WILLIAMS, Senior Program Officer
EUGENE J. CHOI, Program Officer (through May 18, 2007)
IAN M. CAMERON, Associate Program Officer (through May 21, 2007)
MARTA V. HERNANDEZ, Associate Program Officer (as of March 15, 2008)
SUSAN G. CAMPBELL, Administrative Coordinator
MARY G. GORDON, Information Officer
SEKOU O. JACKSON, Senior Project Assistant

Preface

By the end of the Cold War the United States had produced a navy, to a large degree in response to the global challenge posed by the Soviet Union, that was and remains the largest and most powerful navy in the world. This maritime supremacy confers great advantages on the United States in its foreign policy, but it has limitations. The U.S. Navy, while the dominant maritime force, must act in concert with other maritime forces in the quest for an orderly maritime domain. More and more, today's dynamic maritime security landscape also involves such broad-ranging missions as countering global terrorism, providing humanitarian relief for natural disasters, interdicting drug trafficking, and regulating the migration of people. No single navy or nation can do this alone. Security threats in the maritime domain are an important challenge. In today's world 50,000 large ships carry about 80 percent of the world's trade.[1] To offer security in the maritime domain, governments around the world need the capabilities to confront directly such common threats as piracy, smuggling, drug trading, illegal immigration, banditry, human smuggling and slavery, environmental attack, trade disruption, weapons proliferation, and terrorism.[2]

Recognizing this new international security landscape, the former Chief of Naval Operations (CNO) called for a collaborative international approach to maritime security. Initially branded the "1,000-ship Navy,"[3] this concept envi-

[1]VADM John G. Morgan, USN, and RDML Charles W. Martoglio, USN. 2005. "The 1,000-Ship Navy: Global Maritime Network," *U.S. Naval Institute Proceedings,* November, p. 15.

[2]VADM John G. Morgan, USN, and RDML Charles W. Martoglio, USN. 2005. "The 1,000-Ship Navy: Global Maritime Network," *U.S. Naval Institute Proceedings,* November, p. 15.

[3]Chief of Naval Operations (ADM Michael G. Mullen, USN), in remarks delivered to the 17th International Seapower Symposium, Naval War College, Newport, R.I., September 21, 2005.

sioned that U.S. naval forces would partner with "a diverse array of multinational, federal, state, local and private sector entities to ensure freedom of navigation, the flow of commerce, and the protection of ocean resources."[4] Furthermore, said the former CNO, this vision would bring all nations together to build a global maritime network—including the sharing of information that would be available to all participants—that would promote security on the seas and enable global, regional, and national prosperity through international cooperation and coordination. Working toward this vision would often involve developing partnerships with selected nations on a regional or even subregional basis.

Some key components for the 1,000-ship Navy vision to be successful have been identified.[5,6] First, there must be incentives for participating nations to join in such a partnership. Most maritime threats are not global; therefore, the regional and local interests of each country must be accommodated. The operating principle behind the 1,000-ship Navy is that it must satisfy the interests of all participants, many of them held in common. Second, there must be low barriers to entry, both technologically and operationally, to make this truly a coalition of the willing for all nations, even those without formal navies. Third, the partnership should advance security, local and global economic prosperity, and overall cooperation among governments. For example, as the network grows one would expect to see increases in the number of sensors and responders available to monitor and support security in the maritime domain.[7] Finally, building trust among all nations should be an overarching objective of such partnerships and one that will be crucial for such a coalition of the willing to be realized.

Since the CNO's speech in 2005, the U.S. Navy has been actively engaged in working demonstrations of the 1,000-ship Navy concept; in particular, the U.S. Navy has participated in multinational counterpiracy efforts off the coast of East Africa as well as in conducting training with navies in the Gulf of Guinea and Latin America.[8] Also, the U.S. Navy, the U.S. Marine Corps, and the U.S. Coast Guard have jointly released a new maritime strategy document that gives preventing a war the same military priority as winning a war and advocates more

[4]Chief of Naval Operations (ADM Michael G. Mullen, USN) and Commandant of the Marine Corps (Gen Michael W. Hagee, USMC). 2006. *Naval Operations Concept,* Department of the Navy, Washington, D.C.

[5]Christopher P. Cavas. 2006. "The Thousand-Ship Navy," *Armed Forces Journal,* December.

[6]RDML Jeffrey A. Wieringa, USN, Deputy Assistant Secretary of the Navy (International Programs) and Director, Navy International Programs Office, Office of the Secretary of the Navy, "DASN (IP): Role of International Programs in Developing the 1000-Ship Navy," presentation to the committee, January 10, 2007.

[7]VADM John G. Morgan, USN, and RDML Charles W. Martoglio, USN. 2005. "The 1,000 Ship Navy: Global Maritime Network," *U.S. Naval Institute Proceedings,* November.

[8]Chief of Naval Operations (ADM Michael G. Mullen, USN). 2007. *CNO Guidance for 2007: Focus on Execution,* Department of the Navy, Washington, D.C., February 2.

cooperation with foreign fleets.[9] The concept of such maritime partnerships has received positive support from the world's maritime leaders.[10,11]

TERMS OF REFERENCE

At the request of the former Chief of Naval Operations,[12] the Naval Studies Board of the National Research Council conducted a study to examine the technical and operational implications of the 1,000-ship Navy concept as they apply to four levels of cooperative efforts: (1) U.S. Navy, Coast Guard, and merchant shipping only; (2) U.S. naval and maritime assets with others in treaty alliances or analogous arrangements; (3) U.S. naval and maritime assets with ad hoc coalitions (examples to be postulated in the study); and (4) U.S. naval and maritime assets with others than the above that may now be friendly but could potentially be hostile, for special purposes such as deterrence of piracy or other criminal activity. Specifically, for each of these four levels, the study addressed the following tasks:

- Examine previously established models and other possible operational concepts for the four levels of cooperation, to include both the NATO and Interpol models;
- Identify force structure and interoperability needs, to include information sharing and assurance;
- Examine the extent to which sensor technology, information and operational techniques must be held classified; and the utility, advantages and disadvantages of using civilian communications and encryption technologies; and
- Assess potential vulnerabilities and countermeasure susceptibilities to U.S. military forces inherent in the "1,000-ship Navy" concept, and the means to mitigate them.

THE COMMITTEE'S APPROACH

There has been much discussion about whether the U.S.-led initiative 1,000-ship Navy would be widely acceptable to potential partners and would offer the

[9]Department of the Navy, 2007, *A Cooperative Strategy for 21st Century Seapower*, Washington, D.C., October; *Inside the Navy*, 2007, "Document Released: New Maritime Strategy Urges Tighter Ties for Sea Services," October 22.

[10]Chief of Naval Operations (ADM Michael G. Mullen, USN). 2007. *CNO Guidance for 2007: Focus on Execution*, Department of the Navy, Washington, D.C., February 2.

[11]U.S. Naval Institute. 2007. "The Commanders Respond," *U.S. Naval Institute Proceedings*, March.

[12]ADM Michael G. Mullen, USN, Chief of Naval Operations. Letter dated June 29, 2006, to Ralph J. Cicerone, President, National Academy of Sciences.

respect for their national sovereignty that is critically important.[13,14,15] This concern led the committee to a search for alternative terminology for the 1,000-ship Navy and for an appropriate and effective mechanism of leadership and coordination. The committee adopted the term "maritime security partnerships" (MSP) in this report.[16] The technical and operational implications of MSP are addressed in this report, along with the mechanisms of leadership and coordination.

As the study progressed, the committee refined its understanding of the four levels of cooperative efforts for maritime security called out in the terms of reference and discovered that a different organizing principle would be more appropriate given the complexity of the 1,000-ship Navy concept as it is being developed and implemented. The committee's approach was to take into account and build on the ongoing efforts and respond to the spirit of the CNO's request for the study while at the same time addressing the four levels of cooperative efforts for maritime security and the tasks (the four bullet items) in the terms of reference. Discussions with the Deputy Chief of Naval Operations for Information, Plans, and Strategy (N3/N5) at the committee's first meeting encouraged a broader approach to the study—namely, to consider the more important question of how to achieve MSP. This meant going beyond technical and operational support for MSP to the matter of bringing in the wide range of participants called for in the terms of reference. With this in mind the committee turned to a somewhat more complicated set of bilateral and multilateral models of cooperation to address the tasks in the terms of reference. Chapter 1 presents the committee's understanding and assumptions, its approach to the terms of reference, and the organization and content of this report.

The committee[17] was first convened in January 2007. It held additional meetings and site visits over a period of 6 months, both to gather input from the relevant communities and to discuss its findings and recommendations. The agendas of the meetings are summarized below:[18]

- *January 9-10, 2007, in Washington, D.C.* Organizational meeting: Office of the Chief of Naval Operations, Office of the Secretary of the Navy, Headquarters U.S. Coast Guard, Headquarters U.S. Marine Corps, U.S. Maritime Administration, and Office of the Principal Deputy Undersecretary of Defense for Policy briefings on the operational and technical implications of the 1,000-ship Navy.

---

[13]Amy Klamper. 2006. "Traction," *Seapower*, December.

[14]Michael W. Coulter, Deputy Assistant Secretary for Regional Stability, Bureau of Political-Military Affairs, Department of State. Discussion with the committee, February 6, 2007.

[15]CAPT Bruce B. Stubbs, USCG (Ret.). 2007. "Making the 1,000-Ship Navy a Reality," *U.S. Naval Institute Proceedings,* January.

[16]As of this writing, the U.S. government was moving to replace the name "1,000-ship Navy" with "global maritime partnerships" (GMP).

[17]Biographies of its members are provided in Appendix A.

[18]During the course of its study, the committee held meetings in which it received (and discussed) materials that are exempt from release under 5 U.S.C. 552 (b).

- *February 6-7, 2007, in Washington, D.C.* Military Sealift Command; Office of the Under Secretary for Science and Technology in the Department of Homeland Security; Naval Meteorology and Oceanography Command; Office of the Deputy Assistant Secretary for Regional Stability, Bureau of Political-Military Affairs, Department of State; CNO Strategic Studies Group; International Maritime Organization; and the Maersk Line, Ltd., briefings on the global perspective of the 1,000-ship Navy as well as on the policy, operational, and technical implications of the 1,000-ship Navy.
- *March 12, 2007, in Suitland, Maryland.* Site visit to the Office of Naval Intelligence.
- *March 13-14, 2007, in Washington, D.C.* Defense Information Systems Agency, Office of Naval Research, the Royal Navy (U.K.), Office of the Director of National Intelligence, Interpol, Embassy of Singapore, Office of the Assistant Secretary of Defense for Networks and Information Integration/Office of the DOD (Department of Defense) Chief Information Officer, Consortium for Oceanographic Research and Education, and Office of the Chief of Naval Operations briefings on networks and information-sharing needs for and global perspective of the 1,000-ship Navy.
- *March 29-30, 2007, in Naples, Italy.* Site visit to Commander, Naval Forces Europe–Commander, Sixth Fleet, and NATO Component Command, Maritime Naples.
- *April 2-3, 2007, in London, England.* Site visits to Lloyd's of London; International Maritime Organization; Greek Shipping Co-operation Committee; Shell International Trading and Shipping Company, Ltd.; the Royal Navy (U.K.); and Royal United Services Institute for Defence and Security Studies.
- *April 16-19, 2007, in Alexandria, Virginia.* Site visit to Maritime Domain Awareness Data Sharing Community of Interest Pilot Spiral 3 and Maritime Domain Awareness Connectivity Technology Insertion Game Workshop.
- *April 25-26, 2007, in Miami, Florida.* Site visits to Center for Southeastern Tropical Advanced Remote Sensing at the University of Miami, Headquarters of the Seventh Coast Guard District, and the U.S. Southern Command.
- *May 15-17, 2007, in Washington, D.C.* Office of the Chief of Naval Operations; Headquarters U.S. Coast Guard; National Security Council; Office of the Deputy Assistant Secretary for Regional Stability, Bureau of Political-Military Affairs, Department of State; Naval War College; Center for Naval Analyses; Naval Criminal Investigative Service; Joint Interagency Task Force-South; U.S. Pacific Command; Joint Interagency Coordination Group; Joint Interagency Task Force West; Pacific Fleet N5; Indian Embassy; and Chilean Embassy briefings on the global perspective of the 1,000-ship Navy, as well as the policy, operational, and technical implications of the 1,000-ship Navy.
- *May 17, 2007, in Laurel, Maryland.* Site visit to the National Security Agency Information Assurance Directorate.

- *June 12, 2007, in Key West, Florida.* Site visit to the Joint Interagency Task Force-South.
- *June 25-29, 2007, in Woods Hole, Massachusetts.* Committee deliberations and report drafting.

The months between the committee's last meeting and the publication of the report were spent preparing the draft manuscript, gathering additional information, reviewing and responding to the external review comments, editing the report, and conducting the security review needed to produce an unclassified and unrestricted report.

Acknowledgment of Reviewers

This report has been reviewed in draft form by individuals chosen for their diverse perspectives and technical expertise, in accordance with procedures approved by the National Research Council's (NRC's) Report Review Committee. The purpose of this independent review is to provide candid and critical comments that will assist the institution in making its published report as sound as possible and to ensure that the report meets institutional standards for objectivity, evidence, and responsiveness to the study charge. The review comments and draft manuscript remain confidential to protect the integrity of the deliberative process. We wish to thank the following individuals for their review of this report:

Michael Brock, MITRE Corporation,
John D. Christie, LMI,
Archie R. Clemins, ADM, USN (retired), Boise, Idaho,
Robert L. Jervis, Columbia University,
William B. Morgan, Rockville, Maryland,
David A. Richwine, MajGen, USMC (retired), Burke, Virginia,
Harvey Sapolsky, Massachussetts Institute of Technology,
John Tozzi, RADM, USCG (retired), L-3 Communications Systems, and
George M. Whitesides, Harvard University.

Although the reviewers listed above provided many constructive comments and suggestions, they were not asked to endorse the conclusions or recommendations, nor did they see the final draft of the report before its release. The review of this report was overseen by Alexander H. Flax, Potomac, Maryland. Appointed by the National Research Council, he was responsible for making certain that an independent examination of this report was carried out in accordance with institutional procedures and that all review comments were carefully considered. Responsibility for the final content of this report rests entirely with the authoring committee and the institution.

Contents

APPENDIXES

Summary

At the outset of his tenure as Chief of Naval Operations (CNO), ADM Michael G. Mullen, USN, adopted a progressive vision for the peacetime engagement of naval forces—namely, to enhance the stability and security of the maritime environment. He called this vision "the 1,000-ship Navy." To help develop the concept, ADM Mullen asked the Naval Studies Board, under the auspices of the National Research Council, to establish a committee that would examine the technical and operational implications of the 1,000-ship Navy.[1] In response to the emphasis in the study's terms of reference on the sharing of maritime information and on coordinated tactical action to help maintain order on the seas for all concerned, the committee has chosen to call this concept "maritime security partnerships" (MSP).[2]

In addition to discussions with senior naval personnel, combatant commander representatives, and other Department of Defense (DOD) elements, the committee surveyed a broad cross section of international organizations, foreign navies, U.S. government agencies, and private industry to understand the issues, opportunities, and common needs presented by MSP.[3] Some key observations stand out from all the briefings that the committee received during the course of this study:

- Governments of countries other than the United States tend to be con-

[1]ADM Michael G. Mullen, USN, Chief of Naval Operations, in a letter dated June 29, 2006, to Ralph J. Cicerone, President, National Academy of Sciences.

[2]As of this writing, the U.S. government was moving to replace the term "1,000-ship Navy" with "global maritime partnerships" (GMP).

[3]See the summarized agendas of the meetings in the Preface.

cerned much more with the need for information on traditional maritime security concerns—smuggling, poaching, and piracy—rather than information on direct threats of external attack;

- Most representatives of foreign governments and foreign and domestic commercial organizations expressed interest in collaborating on MSP;
- A number of foreign countries and foreign and domestic commercial organizations might find it difficult to cooperate in MSP activities if these activities were under the U.S. Navy, U.S. DOD (and its intelligence community), or even the U.S. federal government; they might be more receptive to collaboration if entities like the State Department or the U.S. Coast Guard (USCG)—along with the International Maritime Organization (IMO), a relevant international entity—played the key role(s);
- The purposes of the partnerships set up under the MSP concept, often regional in scope, are expected to be the maintenance of law and security on the seas for all concerned; and, finally,
- The partnerships will most likely need to offer full protection for proprietary and country- or company-sensitive information.

WHY MARITIME SECURITY PARTNERSHIPS?

Today's interdependent global economy depends on free and uninterrupted use of the sea. The security and welfare of all nations are linked to a regime of law and order at sea that suppresses illicit activities such as drug smuggling and human trafficking and thwarts threats of piracy and terrorism. The U.S. Navy is well positioned to help other maritime forces and organizations maintain an orderly maritime domain. How the U.S. government and in particular the U.S. Navy should organize, operate, and seek to develop relationships with other governments in pursuit of this goal is the subject of this report.

The complexity of the maritime domain and the diversity of interests at stake militate against relatively simple yet all-encompassing solutions, because the problem is much broader in scope than the naval force or forces of any single country or group of countries can deal with. MSP would need participation by many agencies involved in law enforcement, homeland security, and foreign policy. In addition to the foreign militaries, law enforcement agencies, local civil authorities, and the like with which the United States already liaises, commercial and nongovernmental actors—for example, shipping and insurance companies—would also need to be involved. More broadly, the committee envisions emerging maritime security partnerships to be grounded in international agreements like those for air traffic management, other law enforcement enterprises, financial transaction governance, and the safety of life at sea, with the last-mentioned coming under the IMO, an agency of the United Nations.

UNIFYING CONCEPT FOR MSP—TO ACHIEVE MARITIME DOMAIN AWARENESS INFORMATION SHARING

The effort to improve the security of some legitimate and important maritime enterprises is seriously impeded by the lack of adequate maritime security frameworks in many regions of the world. Individuals or groups who want to disrupt trade along these routes by taking advantage of the tradition of anonymity often found at sea can engage in illegal and threatening activities. To adequately surveil all the commercially critical sea lanes, choke points, natural resource locations, and potential smuggling routes and to maintain links to maritime security forces are major challenges. Some of the questions that need to be answered are these: Who will pay for the costs of such systems? Who will create and coordinate the policies behind the surveillance, information exploitation and distribution, and response plans?

The unifying concept for maritime security partnerships is information sharing. Using the vocabulary that has been adopted in the U.S. initiatives responding to the National Strategy for Maritime Security (NSMS), the information to be shared is referred to as maritime domain awareness (MDA).[4] A comprehensive MDA system would permit identification of threatening activities and anomalous behavior. Achieving such a system where it does not now exist—and strengthening it where there is already a foundation—must be viewed as a critical step in building regional partnerships.

It is important to recognize that some regions have established networks to achieve maritime domain awareness by sharing information. For example, the Malacca Strait Security Initiative partnering Singapore, Indonesia, and Malaysia is already operational; the Gulf of Guinea network, still in its formative stage, has generated great interest among potential partners; the Joint Interagency Task Force-South that addresses concerns about drugs and other law enforcement matters in the Caribbean region is functioning effectively. There is a worldwide patchwork of capabilities in support of MDA systems but no overarching MDA architecture. Current arrangements, some of them multilateral, for sharing MDA information constitute an inefficient assortment lacking broad application; exceptions are the IMO-sanctioned Automatic Identification System (AIS) and Long-Range Identification and Tracking (LRIT) reporting systems for commercial ships.

It will take a major effort to coordinate all the existing capabilities, extend them, and disseminate the information on a timely basis to those maritime law enforcement organizations that can take necessary and appropriate action, while still respecting commercial and national sensitivities and proprietary interests.

---

[4]The Department of Homeland Security's 2005 *National Plan to Achieve Maritime Domain Awareness* (Washington, D.C., October, p. 1) defines MDA as "the effective understanding of anything associated with the maritime domain that could impact the security, safety, economy, or environment of the United States."

Mobilizing the U.S. government to assist other nations in creating more comprehensive MDA and building, connecting, and enlisting the capabilities of the maritime law enforcement organizations will probably be a long process that needs continuous work and attention. At the same time, however, this process would build trust and transparency, contribute to global unity and cooperation, and help to prevent conflict.

The committee's investigations and deliberations are the basis for the findings and recommendations it offers. The committee recognizes the need to balance the demands of maritime security with those of other government missions and priorities. It also recognizes that competing priorities, costs, and missions might stand in the way of the implementation of improvements in maritime security. The committee views all of the recommendations as complementary to one another, and once they have been translated into potential actions with specific costs, they need to be prioritized and compared to other investments.

KEY PREREQUISITE FOR MSP—TRUST IN RELATIONSHIPS

The premise behind the MSP concept is that by improving its situational awareness of what is happening in maritime areas of potential importance to its interests, a state directly improves its own security and therefore ought to be willing to share relevant data with those states it perceives to have congruent interests. Relationship building and information sharing during normal times may also mean that in a time of crisis, the state will be able to call on, or access, individuals or information that can address an emerging problem. The ease and trust with which information or individuals can be accessed will be directly related to the success of the state's past relationships.

Three critical elements are needed to achieve local, regional, or global success in establishing new maritime security partnerships or improving existing ones:

- A cadre of trained, proactive specialists, military and civilian, who are able to operate linguistically and culturally in the context of U.S. planning and coordinating functions within the region;
- Secure, persistent, and adequate funding for specific near-term opportunities for expanded military-to-military exchanges; and
- A robust coordinating authority at the highest levels of the U.S. government that can arrange appropriate governance at all levels (see Chapter 4). It could bring disparate program elements in from across the different agencies and ensure a proactive, coordinated effort to overcome regional challenges or meet urgent local needs.

Regional approaches to either specific or general concerns seem to work the best when congruent interests and stakes are clear or the legacies of past habits of cooperation can be reviewed to support present agreements. Often there is near-universal agreement on the desired outcome, especially where there is strong leadership on the part of major powers and the states that are most affected. But for most maritime security purposes, the committee realizes that "all issues are local."

In some cases, relationships that are formed to address a specific immediate issue start out on an ad hoc basis, but if they are successful they become formalized programs over time. The U.S. government's response to the tsunami in the Indian Ocean in late 2006 is an example of an ad hoc relationship that was formed to respond to catastrophic devastation.

Some programs of cooperation do not succeed directly but rather set the stage for later programs under more auspicious political conditions. For example, the USCG's Caribbean support tender has become a "circuit rider" throughout the region, providing training and maintenance support.[5] A unique aspect of this program was the personal cooperative relationships developed by the international crew of approximately 50. The crew comprised the captain and a small cadre of officers and enlisted men from the USCG and individuals from member countries' officer corps and enlisted units, all of whom sailed on the vessel for a year. After they returned to their own countries, many assumed positions of leadership. Competition for assignment of a national to this vessel was great.

Finding: Most information-sharing relationships start out as an individual bilateral agreement between the United States and one other country. The greatest gains in the intermediate term come from expanding bilateral relationships and agreements. In many cases, the base on which to build will be military-to-military relationships that can be expanded to include other groups—military and civilian, government and nongovernment—that are important to the maritime security task.

Recommendation 1: The Chief of Naval Operations, working with the combatant commanders, the Commandant of the Coast Guard, and the Commandant of the Marine Corps,[6] should commit to transforming bilateral relationships into broader, more substantiative and inclusive maritime security partnerships through some or all of the following means:

- Forward presence;
- Increased language and cultural awareness;

[5]The Caribbean support tender is a U.S. ship dedicated to promoting cooperation with partner nations by visiting countries to conduct maritime training, maintenance assistance, and logistics support.

[6]The identification of specific officers and offices in the government with specific recommended actions is intended to reflect those most closely aligned in terms of the existing structures of organizational responsibilities.

- Expeditionary training teams;
- Ongoing analysis of gaps in capacity with plans for follow-up capacity-building steps;
- Tools and resources appropriate for the particular geography of an area—for example, shallow draft vessels such as the HSV-2 *Swift* rather than larger and deeper draft combat vessels;
- Maritime domain awareness—information-sharing systems that will eventually be expandable to include both unclassified and classified information; and
- Funding for Phase Zero.[7]

Exercises and exchanges are a fundamental vehicle for building trust, which will lead to nation-to-nation cooperation. Information sharing can be facilitated through combatant commander (COCOM) maritime operations centers and headquarters to develop awareness and to develop relationships with partner nations. Training for cooperation lends itself readily to war gaming, another effective vehicle. Face-to-face gaming with foreign partners will address the issues of cooperation before matters reach the point of actual engagement.

The instruments of operational cooperation range from equipment and systems to the training of U.S. and partner nation personnel in the COCOM's area of responsibility. Clearly, all of these partners must have the equipment and software systems to interface with an information-sharing database and/or to feed the database. Data standards for sharing must be developed.

Finding: The continued training of U.S. and partner nation personnel in a maritime security partnership is critical to long-term success and to building the relationships and trust that eventually result in the establishment of maritime security partnerships with as many countries as possible.

Recommendation 2: To educate and train U.S. and partner nation personnel so that they can support and extend maritime security partnerships, the Chief of Naval Operations should:

- With the active support of the leaders of the Marine Corps and the Coast Guard, ask the combatant commanders to support and extend maritime security partnerships through continued and even expanded formal educational and bilateral/multilateral training exercises for these personnel;
- Require that maritime security training become a significant part of the core curriculum at every level of professional education for maritime service;

[7]The traditional four phases of a military campaign identified in Joint publications are deter/engage, seize initiative, decisive operations, and transition. Phase Zero encompasses all activities prior to the beginning of Phase I—that is, everything that can be done to prevent conflicts from developing in the first place.

- Adopt as a critical long-term goal the broadening of participation in maritime professional education to ensure representation from all of the relevant U.S. civilian and military agencies; and
- Cooperate with the Secretary of the Navy and join in the present Coast Guard plan under the Department of Homeland Security to design and fund an institute of maritime studies that would encompass specialized studies in maritime security within the framework of an existing university program.

Critical for the longer-term ability of the CNO to implement MSP will be the establishment within the maritime services of a clear professional career track for officers and civilian officials with wide-reaching international expertise and experience. Appropriate models are the foreign area officer (FAO) programs of the Army and, to a lesser extent, the Air Force and the Marine Corps.

Finding: There appears to be a shortage of qualified FAOs within the U.S. naval services. Such FAOs could provide invaluable aid in developing the capabilities of regional maritime security forces that would allow them to move their countries toward participation in regional and, later, global maritime information sharing.

Recommendation 3: The Chief of Naval Operations should mandate the expansion of a robust foreign area officer (FAO) program within the Navy to meet the needs of staffing and expanding maritime security partnerships. In addition, the Commandant of the Coast Guard should establish an FAO program and the Commandant of the Marine Corps should expand its present limited FAO program for the development of bilateral and multilateral relationships.

The law enforcement authority and legal skills that would be needed to carry out countersmuggling and counterterrorist activities in coastal waters do not usually exist aboard naval vessels. Naval vessels engaged in counter-drug-smuggling missions carry USCG law enforcement detachments (LEDETs) that actually board intercepted vessels that are suspected of smuggling drugs and, if needed, arrest their crews. Using personnel from the Naval Criminal Investigative Service (NCIS) or other law enforcement personnel could be equally effective, but additional training and equipment might be needed to gain ship boarding capabilities as well as to clarify the legal authorization.

Finding: The inclusion of U.S. Coast Guard personnel, the Naval Criminal Investigative Service, or other law enforcement detachments or personnel on selected U.S. Navy ships could extend U.S. capabilities to respond to suspected smuggling or terrorist activities.

Recommendation 4: The Chief of Naval Operations should ask the Coast Guard, the Naval Criminal Investigative Service, or another law enforcement entity to provide legal personnel for selected U.S. Navy ships.

In order to realize theater engagement or Navy MSP goals, the USCG could be asked to forward-deploy additional vessels to specific areas of the world. These vessels would work for the COCOMs on missions accepted by the USCG. For instance, low-end USCG vessels might be the appropriate maritime component command for a military operation. The USCG's "sovereignty expertise" might be the right answer for the Navy/COCOM, allowing them to gain access that they could not otherwise obtain. Such actions could pave the way for greater trust and cooperation between countries, including between their military counterparts.

Finding: The forward deployment of U.S. Coast Guard vessels can enhance and strengthen the engagement activities and thus increase the number of partnerships.

Recommendation 5: The Chief of Naval Operations should ask the Commandant of the Coast Guard to forward-deploy Coast Guard cutters to locations that offer opportunities for the joint enforcement of maritime security. These cutters would help to attain Navy and combatant commander engagement goals and would be the correct security assets to employ to meet theater cooperation goals.

Relatively speaking, the total effort required to expand the scope and depth of MSP is not large. Indeed, some of the overall funding can come from direct or in-kind contributions of the strategic partners themselves. MSP are based on the win-win concept—that is, they are of benefit both to relationships and to the flows of activity and information that sustain them. But at least for the initial period, the 1,000-ship Navy concept requires the Office of Management and Budget to scrutinize Navy programs and budgets not only to identify programs but also to include the funding needed for implementation of the MSP.

Finding: Secure, continuing funding is a key ingredient for sustaining and deepening maritime security partnerships.

Recommendation 6: To sustain and deepen maritime security partnerships (MSP) and to make such programs robust and stable, the Chief of Naval Operations should:

- Establish and assign to a specific office the coordination authority for programs and budgets for MSP in the Navy, throughout the Department of Defense (DOD), and across the federal agencies. This should include enhanced opportunities for professional education and for the necessary equipment and support services;
- Request that the Defense Security Cooperation Agency work with the State Department to significantly enhance the portfolio of international military education and training funds (e.g., those under Sections 1206 and 1208 of the National Defense Authorization Act of 2006, and COCOM Initiative Funds) for

countries deemed key for MSP development. This activity—the implementation of a network of MSP—should also set a high priority on the institutionalization of an international legal training program;

- Task the Navy's International Programs Office to place high priority on funding the transfer of equipment, software, and services to support and intensify existing MSP and to develop new bilateral and multilateral MSP;
- Together with the appropriate officials at the State Department and other agency partners in MSP, request more funds for use by the maritime services, the State Department, and other relevant government agencies for training and support of MSP initiatives or for activities at the International Maritime Organization and other relevant international organizations and multilateral frameworks to maintain and expand information-sharing programs and protocols;
- Propose to the appropriate parts of DOD the setting aside of a portion of research, development, test, and evaluation funds over the next 5 years to be committed under the Office of Manpower and Personnel guidance to the specific goal of improving technologies and techniques for easy, reliable information sharing and the continuous availability of common maritime operational pictures on as broad a basis as possible. These would subsume but go beyond the already programmed funding for MDA only that is now appropriated to the Office of Naval Research (see Chapter 3).

KEY ENABLER FOR MSP—SYSTEMS AND TECHNOLOGIES FOR INFORMATION SHARING

Information sharing is the key to building trust and provides a basis for decisions and actions. The resulting transparency arguably contributes to the shared security interests of the United States and its current and potential partners. More specifically, in the committee's view:

- Maritime security around the globe will be advanced by strengthening existing partnerships and building new ones, with MDA information sharing a key enabler.
- It is in the interest of both the United States and its partner nations to extend information sharing as widely as possible within regions, subregions, and beyond, noting that threats to security typically cross regional and subregional boundaries.
- The related objectives of extending reach and maximizing inclusiveness suggest that both the information to be shared and the system architecture for its sharing must exhibit certain attributes: a focus on the sharing of unclassified information[8] and on the use of commercial, Internet-based sharing mechanisms.

[8]Pursuant to Executive Order 12958, "classified information" refers to official information that has been determined to require protection against unauthorized disclosure in the interest of national security and that has been so designated. "Unclassified information" refers to information that has not

- Beyond information sharing—viewed here as having value in and of itself and as a building block when forming new partnerships—there is, of course, the matter of taking endgame action to interdict illegal or threatening activities. In this regard, strengthened information sharing can be expected to enhance coordination among maritime partners.
- Effective information-sharing architectures and systems are operating today at the classified and unclassified levels.

The committee reviewed baseline U.S. operational and developmental information-sharing systems and related planning and research efforts, emphasizing the technical engineering mechanisms to advance MSP initiatives with nontraditional partners. The committee's findings and recommendations related to systems and technologies for MDA information sharing are presented below.

Finding: Effective information-sharing architectures and systems are operating today at the classified and unclassified levels. Navy and combatant commander (COCOM) efforts with nontraditional partners rely on the Internet model and use of commercial products, including for information protection. However, there is no known, concerted effort to ensure that the Navy's technical efforts are fully connected to or fully leveraged by COCOM or other initiatives. This less than satisfactory level of effort could lead to interoperability problems or could distract COCOM or other operational elements from their mission focus.

Recommendation 7: The Chief of Naval Operations and the Secretary of the Navy should jointly charter and fund an activity, led by the Deputy Chief of Naval Operations for Communication Networks (N6) and supported by appropriate laboratory/system command/program executive office (PEO) expertise, to provide responsive, dedicated technical support across the full range of interagency initiatives for the design, engineering, and fielding of information technology (IT) infrastructure that would enable information sharing for maritime security.

The activity called for by this recommendation would support combatant commanders, Navy operational elements, other U.S. government organizations, and—through them—foreign partners. It would:

- Develop information-sharing design templates and a catalog of imple-

been determined to warrant classification; however, some unclassified information may be approved for public release, whereas other such information, such as International Traffic in Arms Regulations information, may not. Some maritime information that does not pertain to U.S. national security, such as Automatic Identification System reports, can be considered as publicly available and can therefore be freely shared (subject only to constraints imposed by international agreements, such as IMO, as opposed to U.S. policy). When referring to such information, the U.S. Navy has coined the term "not classified," apparently to convey the notion of useful information sharing without the potential complexities of codified protection requirements. The term "unclassified," as used in this report, is viewed as encompassing "not classified" information.

menting products (these might be different for partners within the U.S. government, those within formal alliances, those in ad hoc coalitions, and those with whom information-sharing arrangements are independent of formal alliance or coalition agreements);

- Assemble and engineer starter kits in support of operational initiatives;
- Include available tools for communications, collaboration, and consultation within the broader design template, MSP catalogs, and the starter kits effort outlined above;
- Explore potential value-added upgrades for the future and recommend upgrades and backward compatibility approaches;
- Emphasize the sharing of unclassified MDA information, suitably protected to respect privacy and law enforcement concerns; and
- Perform an end-to-end information protection analysis to ensure that the protection meets the expectations of the partners for the several networks in operation or under development.

These measures would increase coherence among inherently distributed regional or subregional initiatives.

Finding: There is a range of technical options for improved ocean surveillance, some of them near term, that should be less costly than fielding large, new sensor systems. Some of them exploit data from a growing inventory of commercial remote-imaging sensors and satellites, others entail maritime-directed upgrades to existing over-the-horizon radars and/or national reconnaissance systems, and, finally, still others involve coastal radar surveillance of the near-in waters of partner states.

Recommendation 8: The Chief of Naval Operations (CNO) should direct the Director of Naval Intelligence (N2) and the Deputy Chief of Naval Operations for Communication Networks (N6), and the Assistant Secretary of the Navy for Research, Development, and Acquisition should direct the appropriate laboratories, system commands, and program executive offices to increase their efforts to investigate, analyze, and help field, if appropriate, the most cost-effective combinations of capability across the potentially promising approaches to persistent, improved broad ocean surveillance that are identified in Chapter 3. To facilitate this initiative, the CNO should (1) seek a higher level of representation at the National Reconnaissance Office, where decisions are made on U.S. sensor performance goals, and (2) leverage its newly expanded role in the Office of the Director of National Intelligence (ODNI) to encourage the inclusion of maritime surveillance features in the next generation of commercial remote sensors from which the ODNI expects the agencies, particularly the nongovernmental agencies, to contract for products.

Finding: In many parts of the world, U.S. naval component commanders are well positioned to encourage coastal nations to improve their own maritime surveillance capabilities. To this end there are some relatively low-cost, high-payoff improvements for which the Navy could provide not only technical assistance (an example would be the selection and siting of coastal radars) but also material assistance by such means as the Section 1206 funding mechanism.[9] In some places such programs are well under way, but many more opportunities could be productively pursued.

Recommendation 9: The Chief of Naval Operations, in coordination with the combatant commanders, should direct the Director of Naval Intelligence (N2) and the Deputy Chief of Naval Operations for Communication Networks (N6), and the Assistant Secretary of the Navy for Research, Development, and Acquisition should direct the appropriate laboratories, system commands, and program executive offices to ensure that naval component commanders have the appropriate expertise and other assets to facilitate an outreach program to coastal states that would benefit from improved maritime surveillance capabilities.

Finding: Research and demonstration programs sponsored by various agencies have produced good work that addresses some of the technology gaps in the current analysis and fusion of maritime domain awareness information. Much of the technology being developed to analyze and fuse data on maritime entities is in the early stage, in prototype form. However, as reflected in Navy efforts ongoing as of this writing, there are commercial off-the-shelf and potentially releasable government off-the-shelf analysis and fusion tools and software that offer early, useful capabilities for maritime security partnerships.

Recommendation 10: To leverage analysis and fusion technology and tools, the Chief of Naval Operations should assign the Deputy Chief of Naval Operations for Communication Networks (N6) (along with the relevant laboratories and systems commands) to take responsibility for maritime domain awareness-related analyze-and-fuse technologies, either for their short-term application as part of a starter kit (in releasable government or commercial off-the-shelf form) or for longer-term advanced research with identification of transition opportunities. Given that these efforts are of long-term importance, independent of the purposes of current supplementals, the Deputy Chief of Naval Operations for Resources, Requirements, and Assessments (N8) should work on funding maritime domain awareness efforts in the mainstream of the Navy budget.

---

[9] Section 1206 funding, named for the section of the 2006 National Defense Authorization Act that authorizes it, is designed to help other countries build capacity within their national military forces. The authority allows DOD, in consultation with the State Department, to spend up to $200 million a year to help other countries.

Finding: There is a need—unsatisfied today—for a systematic, analytical approach to optimizing the design of the end-to-end system for the collection and analysis of maritime security information and its follow-up. Satisfying this need would require a range of technical support from the Department of Defense and interagency arena to foreign partners.

Recommendation 11: The Chief of Naval Operations and the Secretary of the Navy should jointly propose a Navy-led and Navy-housed executive agent on the technical aspects of an information-sharing system for the U.S. interagency maritime security partnerships initiative. This agent would provide systems engineering and operations analysis resources with technical support to International Maritime Organization initiatives. This mission-driven, enterprise-level systems engineering and analysis capability would be an extension of the Maritime Domain Awareness Executive Agent role already assigned to the Navy by the Department of Defense. It would support not only the U.S. elements but also, under the auspices of ongoing initiatives, its foreign partners.

IMPLEMENTATION STRATEGY FOR MSP—ROLES AND RESPONSIBILITIES ACROSS U.S. GOVERNMENT AGENCIES

The trend during the past two decades toward globalization in the exploitation of natural resources and in the manufacturing sector has meant an increasing need for maritime transport. This need in turn results in growing coastal trade, transoceanic commerce, shipbuilding, port expansion, fuel consumption, and competition for offshore resources—including fish stocks—all of which have a significant impact on national and international governance related to maritime safety, control, and security. The governance burden, especially as regards security, is already straining U.S. resources for protecting the country's own waters and ports. It is time to act on this understanding and prepare the nation and its prospective partners to deal with the growing task of maritime governance.

Establishing a regime such as that implied for MSP is an extensive and exceedingly complex task that needs to involve departments and agencies across the U.S. government. It needs to engage other participants in ways that transcend formal military and political alliances, and it needs to be seen by other countries not as a U.S. military initiative but as a way of fostering law and order at sea and thus the security of all participants. It is not clear that the existing Maritime Security Policy Coordinating Committee—despite some positive steps at the policy level—has adequate authorities or mechanisms to fully realize MSP objectives as part of the national strategy. The situation bears a strong resemblance to the situation that faced the nation with respect to air transportation before the establishment of the Federal Aviation Administration and the International Civil Aviation Organization.

Finding: The Chief of Naval Operations' initial 1,000-ship Navy concept has become a much larger concept of maritime security partnerships, attracting much international recognition and interest. It has grown beyond a U.S. Navy initiative into a critical matter for all agencies of the U.S. government that deal with international maritime relationships and trade.

Recommendation 12: The Chief of Naval Operations should recommend the appointment of an independent third party such as a presidential commission on maritime security governance tasked to recommend ways of strengthening the nation's maritime security policy, to define the roles and responsibilities of various U.S. government agencies and departments to better implement maritime security partnerships both domestically and internationally, and to move forward as suggested in the 11 other recommendations of this report.

1

Introduction: Creating Maritime Security Partnerships in the Twenty-First Century

BACKGROUND

Recognizing the new international security landscape following the end of the Cold War and after the terrorist attack on the United States on 9/11/2001, the Chief of Naval Operations (CNO), ADM Michael Mullen, USN, called for a collaborative international approach to maritime security.[1] Initially branded the "1,000-ship Navy," this concept envisioned that U.S. naval forces would partner with "a diverse array of multinational, federal, state, local, and private sector entities to ensure freedom of navigation, the flow of commerce, and the protection of ocean resources." Furthermore, this concept would bring all nations together to build a global maritime network—including the sharing of information among all participants—that would promote security on the seas and enable global, regional, and national prosperity through international cooperation.

In response to a request from the former CNO,[2] the Naval Studies Board of the National Research Council established the Committee on the "1,000-Ship Navy"—A Distributed and Global Maritime Network for the purpose of conducting a study to examine the technical and operational implications of the 1,000-ship Navy concept. The terms of reference for the study, the committee's understanding and assumptions and its approach to addressing the terms of reference, and the organization and content of this report are outlined below.

---

[1]Chief of Naval Operations (ADM Michael G. Mullen, USN), in remarks delivered at the 17th International Seapower Symposium, Naval War College, Newport, R.I., September 21, 2005.

[2]ADM Michael G. Mullen, USN, Chief of Naval Operations, in a letter dated June 29, 2006, to Ralph J. Cicerone, President, National Academy of Sciences.

TERMS OF REFERENCE

Conduct a study to examine the technical and operational implications of the 1,000-ship Navy concept as they apply to four levels of cooperative effort: (1) U.S. Navy, Coast Guard, and merchant shipping only; (2) U.S. naval and maritime assets with others in treaty alliances or analogous arrangements; (3) U.S. naval and maritime assets with ad hoc coalitions (examples to be postulated in the study); and (4) U.S. naval and maritime assets with others than the above that may now be friendly but could potentially be hostile, for special purposes such as deterrence of piracy or other criminal activity. Specifically, for each of these four levels, the study will:

- Examine previously established models and other possible operational concepts for the four levels of cooperation, to include both the NATO and Interpol models;
- Identify force structure and interoperability needs, to include information sharing and assurance;
- Examine the extent to which sensor technology, information and operational techniques must be held classified; and the utility, advantages and disadvantages of using civilian communications and encryption technologies; and
- Assess potential vulnerabilities and countermeasure susceptibilities to U.S. military forces inherent in the "1,000-ship Navy" concept, and the means to mitigate them.

THE COMMITTEE'S UNDERSTANDING AND ASSUMPTIONS

As the committee heard from various parties during its work, it became clear that ADM Mullen's concept went well beyond cooperation with the navies of the world and put a premium on the sharing of information relevant for maritime domain awareness (MDA) rather than just having more ships in a literal sense. As a result, the committee began to use the term "maritime security partnerships" (MSP) for the purposes of this report. In addition, the committee came to understand that the U.S. government appears not to be well enough organized to pursue the MSP program at this point and must be particularly attentive to the sensitivities of the countries it wishes to enlist as partners. The United States is not popular in many places around the world, and some of its detractors think that any program it proposes is nothing but an attempt to extend its hegemony. They also fear that the United States is merely seeking intelligence for itself and will not share the information picked up by an MDA system. This prejudice can be overcome by making clear that MSP is not simply an extension of intelligence operations but is a real effort to bring stability and prosperity. The United States can at times appear to be too obsessed with terrorism and nuclear proliferation; this can be overcome by showing a real concern for local problems—for example, fisheries protection. The United States also has to make clear that it is not concerned just with its own homeland defense but instead wants a comprehensive MDA system. Finally, the United States should deny that it needs to patrol the whole of the world's oceans but does not have enough ships for the purpose—that is, that the MSP program is really meant for its own security.

The committee came to understand that the goals of MSP are to foster dependable expectations of security and peaceful development and the ability to act in concert against common security challenges. Effective MSP would enable partner nations to act locally in their own self-interest, especially to protect national sovereignty, and in the general interest of law and order on the seas. The CNO identified some key components of the 1,000-ship Navy vision if it is to be successful:

- First, there must be incentives for participating nations to join in such a partnership. Since not all maritime threats are global, the regional and local interests of each country must be considered. The principle behind the 1,000-ship Navy is that it must serve every participant's interests.
- Second, there must be low technological and operational barriers to entry for all nations (even those without formal navies) to achieve the broad participation needed to attain the goals.
- Third, by advancing security, MSP should improve economic efficiency and social cohesion.
- Finally, building trust among all nations, even those that have not traditionally been friendly, should be the overarching objective of such a partnership and will be crucial for realizing a coalition of the willing.

To be successful, the activities of MSP must be conducted within the framework of international maritime laws and conventions. In practice, the vast majority of international agreements are bilateral. The aim of MSP is to build on those bilateral arrangements to bring about cooperative action, initially on a regional level and then on the global level, within the international framework.

Understanding the Current Situation in the Maritime Domain Today

On the one hand, the world today is continuing to grow its economy. Most of the products of the resulting global trade travel by sea (see Appendix B for a detailed discussion of the sea lanes of commerce). On the other hand, there are gaps in governance of the maritime domain that permit outlaws to pursue their political or criminal ends. The globalization of transportation, information, and finance facilitates the outlaws' ability to cross borders and to exploit seams of lawlessness within and between countries. While most countries have armed forces and train them to defend against aggression by neighbors and to protect their territories, defense and law enforcement organizations around the world need to develop security relationships and capabilities to deal with nonstate political and criminal elements. Maritime forces, which have a long tradition of cooperating with neighbors, international organizations, and legitimate merchant shipping for protection and safety at sea, may see the utility of partnerships to protect their own resources and cope with unlawful activities.

The predominant challenges in the maritime domain today come from a range of hostile actions by nonstate actors, from stealing fish to smuggling drugs, people (including both illegal immigrants and slaves), and weapons of mass destruction, to piracy. There is also the possibility that extremists or other insurgents and terrorists may attack at sea.[3]

These nonstate actors respect no boundaries. They cross them easily (including by sea) to carry out their business. Pirates pose threats to general merchant shipping, particularly in straits or from the coastal areas of undergoverned states like Somalia. Smugglers and fishery poachers might be thought of as "evaders"—that is, they want to evade law enforcement authorities. Pirates, insurgents, and terrorists are attackers. Defending national resources and sovereignty against both evaders and attackers who use the sea requires intelligence about such activities both at sea and ashore. This in turn requires cooperation among military and law enforcement forces, both within countries and between countries.

There is a great need to have a picture as comprehensive as possible of all activities that affect the maritime domain. From gathering tips ashore about evading or attacking activities to gaining a picture of normal activities and the routine reporting of both legal and illegal traffic at sea—all would set the stage for identifying anomalies in the traffic and taking appropriate action, much as does the Joint Interagency Task Force-South (JIATF-S) for drug traffic in the Caribbean.

The Need for Maritime Security Partnerships

Attackers and evaders challenge defense and law enforcement in the maritime domain. Beyond self-defense and pursuit of pirates, the authorities that deal with attackers and evaders are generally confined to their national territories or, in the case of maritime law enforcement authorities, to their own territorial waters, which range from the local waters of a country (12-mile zone, 24-mile contiguous zone, exclusive economic zone [EEZ], or a continental shelf) to the boundaries with neighboring country waters. Large portions of the sea, such as the seas of

---

[3]The terrorist attacks at sea so far have been the attack on the USS *Cole* in Aden Harbor; the attack on the oil tanker MV *Limburg* as it awaited a pilot before entering the port in Mukkala, Yemen; the sinking of a Filipino ferry boat; a rocket attack on U.S. amphibious ships in the Jordanian port of Aqaba; and attacks by the Tamil Tigers (Liberation Tigers of Tamil Eelam) members of the long-standing rebellion in and around Sri Lanka. Also, note that in southeast Asia, particularly in the Indonesian and Filipino archipelagos, terrorists (Abu Sayyaf and Jemaah Islamiah) move people and supplies by sea from island to island. The Naval Studies Board recently conducted a study on the role of naval forces in the global war on terror (see National Research Council, 2007, *The Role of Naval Forces in the Global War on Terror: Abbreviated Version*, The National Academies Press, Washington, D.C.).

Indonesia and the Philippines, are archipelagic. They constitute national territory but are extremely difficult to police thoroughly. Straits and other confined seas (e.g., the Mediterranean) of importance to transit (i.e., they bear heavy merchant traffic) pose other problems of law enforcement and may require country-to-country cooperation if the nonstate threats are significant. In addition, the countries may be concerned about threats approaching from a distance—for example, the United States is concerned about terrorists approaching from across the Atlantic. Altogether, the defense and law enforcement activities of various countries are manifestations of their sovereignty, but at the same time the authorities are largely confined to their national territory. Clarifying authority and extending it to act against attackers and evaders at sea through bilateral and regional arrangements, consistent with international laws and conventions, is an essential goal of MSP.

For the purposes of MSP, one is not just talking about national navies. The enforcers include anything that floats, flies out to sea as part of detection and enforcement, or supports both boats and aircraft from the shore (including port authorities, radar stations, and so on). The law enforcement authorities may include national governments and their security and defense ministries (one such entity is the Guardia Finanzia in Italy), including port authorities; coastal patrols (12-mile and 24-mile zones); capabilities that extend out to the 200-mile limit for EEZ protection; and the more distant warding off of perceived threats (e.g., the international maritime interception operation such as the one in the Persian Gulf or the Operation Active Endeavor of the North Atlantic Treaty Organization (NATO) in the Mediterranean, and so on).

Organizations of this kind may have different names depending on the country and its history. They include navies, coast guards, customs, harbor police, and any other authorities that float on the water or conduct surveillance over the waters. Many countries, and particularly those engaged heavily in international commerce, have the capabilities to police their own waters. Many cooperate in the current patchwork of international cooperation for regulating maritime traffic. As much as three-quarters of the world may be adequately governed in this respect. But there are countries and areas, particularly in Africa, with offshore fisheries and where international trade is growing, that suffer from inadequate governance. Particularly neglected in these countries is the enforcement of laws in the maritime domain. Such countries need help.

Those needing help with protection are the countries, their borders and coastlines, their fisheries, and the general merchant marine—all of which will have a stake in cooperating with the protector organization to be able to continue their peaceful pursuits. Active patrols and enforcement responses are also likely to deter those who might be contemplating unlawful activities in the various maritime areas.

Altogether, these rather scattered threats are ubiquitous and mobile and able to cross borders (most are evaders, not attackers). International cooperation is needed to ensure the protection of the maritime domain against these threats and unlawful activities. At a minimum, information must be shared between one sovereign state and another to facilitate the pursuit of lawbreakers. Two or more neighboring countries may form joint or coordinated patrols. Port authorities around the world are in communication about ships that break regulations by, for example, spreading pollution. Regional and global information sharing is essential for dealing with these challenges.

Understanding the Various Levels of Cooperation for Governance of the Maritime Domain

The study's terms of reference called for the committee to examine the technical and operational implications of the 1,000-ship Navy concept as they apply to four levels of cooperative efforts: (1) U.S. Navy, Coast Guard (USCG), and merchant shipping only; (2) U.S. naval and maritime assets with others in treaty alliances or analogous arrangements; (3) U.S. naval and maritime assets with ad hoc coalitions (examples to be postulated in the study); and (4) U.S. naval and maritime assets with others than the above that may now be friendly but could be hostile, for special purposes such as deterrence of piracy or other criminal activities.

As the committee went about its task by gathering information from representatives of the U.S. Navy, the USCG, merchant shipping, foreign countries, and other organizations—government and nongovernment, including industry, both domestic and international—it came to understand that the maritime domain is not an ungoverned space, particularly because it extends along coastlines. There is already a good deal of regulation of ships (see the rules set forth in the United Nations Convention on the Law of the Sea [UNCLOS] and by the International Maritime Organization [IMO]) and much cooperation between various maritime authorities locally, regionally, and globally. The United States has long been promoting cooperation, starting during the Cold War and continuing afterwards, and including, in the case of the Navy, troop and vessel deployments overseas. Navies and other maritime organizations have natural relations with one another, given their association with the sea. There is also a lot of cooperation among legitimate seagoing business entities—merchant marines, fishing fleets, and so on. It tends to be piecemeal, however, and merchant marines are especially sensitive to protecting their competitive positions in trade.

There are several levels of sophistication and modernity in the countries and their maritime organizations (navies and other organizations involved in law enforcement in territorial waters) that would be involved in MSP, and generally there is close cooperation between them.

- At the high end are the navies that can venture out globally; the committee assumes their territorial capabilities are as good. Most are close allies of the United States, with which they have long-standing cooperation.

• There are other very capable navies that do not roam the world (except perhaps on show-the-flag visits) but generally stay in their own regions, which presumably contributes to the general stability of those regions and also supplements their territorial defenses. Some are close allies of the United States; others have tenuous relations in the maritime domain. The MSP initiative offers a way to engage these countries in cooperative security endeavors.

• Other reasonably developed countries may have capable coastal navies and maritime enforcement organizations that police and protect their own waters. Their participation in MSP would be local and in nearby international waters, and they would benefit from information sharing and coordination with their neighbors for these purposes.

• Finally, there are those underdeveloped countries that have limited or no capabilities, even for coastal patrols. These are countries that have less control of their maritime domains but are perhaps responsible for the greater part of unlawful activities. If they are to participate in MSP, they would need more capable vessels and support from outside for information capabilities.

All of these levels of international cooperation require MDA to function and enforcers able to respond in a timely way, as appropriate. For the system to work, they must share, as well as consume, information. Many, if not most, countries want to know the location of all the ships within their jurisdictions, particularly those heading in their direction or already in their sovereign waters. It may be that the best way to make MSP a reality will be to gather, process, and then arrange to share this information with those who need it in order to conduct enforcement actions.

A critical aspect of law enforcement is interceptions and boardings, which in turn bring up a very important concern for MSP, which is respect for sovereignty and for ensuring that actions taken by any of the maritime entities are legal. There is a certain tension between freedom of the seas and the maintenance of lawful order in sovereign waters. Outside sovereign waters, there are few restraints on the passage of vessels. The right of innocent passage—which the United States especially defends given the ubiquity of its Navy in its deployments around the world—is central. As discussed in Chapter 3, the right to stop and board a merchant ship is restricted.

Another challenge is building the capabilities of the less capable states, especially those in sub-Saharan Africa, where the United States has become especially worried about the security of expanding oil production and shipment from the Gulf of Guinea area. The near-lawless coasts of East Africa are also of concern. Boat people are making their way from West Africa to Europe through way stations like the Cape Verde Islands. The United States and the other advanced nations are going to have to convince the developing countries to give these maritime security efforts priority along with efforts to achieve economic and social development. The costs of building and sustaining the countries' capabilities in maritime law enforcement and connecting them to an MDA information system will have to be underwritten. The insufficiency of governance in many of these countries and their lack of financial resources present substantial difficulties. The United States would also have to reset priorities in its always constrained military assistance budget if it is to support the effort.

Considerations in Improving Law Enforcement in the Maritime Domain

In summarizing its understanding of the scope of MSP, the committee notes that the top layer is the maritime domain—the seawaters of the earth (but not the rivers or internal bodies of water). Then, given that both legal and illegal water-borne traffic crosses boundaries (except, perhaps, in the case of countries with long coastlines, like the United States or India, where there is much intracoastal trade traffic), the second layer comprises the many nations that front the maritime domain. It is this layer that must take coordinated action against attackers and evaders. In arranging a system of partnerships for surveillance, information-sharing, and law enforcement, there are many technical problems to solve.[4] The third layer involves cooperation in enforcement. The three layers of cooperation are shown in Figure 1.1.

In these three layers several considerations are at play:

- *Location of threats.* Threats are everywhere. They may come from the local ports and country waters or from anywhere around the world. This consideration takes into account that the greater part of the world maritime trade moves all over the world. Even fishing vessels may operate well beyond a country's territorial waters.
- *Extent of regulation.* There is some confusion over this consideration—namely, is the maritime domain an ungoverned space, an anarchic space, or a regulated space? Much of international law was developed to govern the maritime domain, but the means of enforcement are often limited. In general, the massive world trade that moves by sea is hardly ever disturbed, even the movement of oil out through the Persian Gulf and the Strait of Hormuz. Piracy currently seems to be limited to particular areas and so far is a minor irritant. Smuggling in its various forms goes on as it has throughout history. So far terrorists have hardly struck at sea (their three recent attempts have been in ports or at the entrance to them). But the maritime space is vast and the surveillance poor except in the close approaches to major ports. One way to reduce the space in which trouble could

[4]See National Research Council, 2000, *Network-Centric Naval Forces: A Transition Strategy for Enhancing Operational Capabilities,* National Academy Press, Washington, D.C.; National Research Council, 2005, *FORCEnet Implementation Strategy,* The National Academies Press, Washington, D.C.; National Research Council, 2006, *C4ISR for Future Naval Strike Groups,* The National Academies Press, Washington, D.C.; National Research Council, 2007, *Distributed Remote Sensing for Naval Undersea Warfare: Abbreviated Version,* The National Academies Press, Washington, D.C.

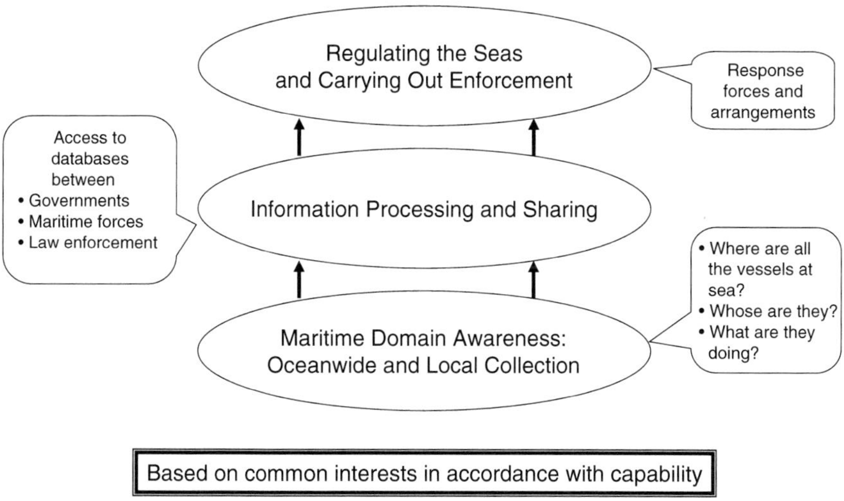

FIGURE 1.1 Maintaining order in the maritime domain.

occur is to fill the gaps in regulations through international efforts, especially through the IMO.

- *Capabilities of individual countries.* The third set of considerations varies from highly capable governments with excellent maritime capabilities all the way to not very capable governments with poor or nonexistent maritime capabilities. While surveillance and information-sharing capabilities can be made available worldwide, it is the countries themselves that are ultimately responsible for moving about on the surface of the seas to carry out enforcement. One objective of MSP is to increase the capability and coordination of these enforcers as well as their numbers to achieve greater coverage in the areas where attackers and evaders operate.
- *Depth of information.* The fourth set of considerations ranges from maximum MDA, which would entail knowing the location of every vessel in the world, down to the specific cases of stopping ships and boarding them for inspections as circumstances admit. There is a substantial element of deterrence to be realized from having a total system in place and recognized as such. Those who wish to take advantage of the vastness of the seas to conduct their nefarious activities might think again if they risked being detected and intercepted even across sovereign boundaries. As seen in the Strait of Malacca and the Strait of Singapore, where patrols have been stepped up, there has already been a reduction in the incidence of piracy.
- *Severity of threat.* A threat that is carried out could have global consequences or local consequences. Terrorists might seize a merchant ship and load

a nuclear weapon on it to be fired when in range, or they might poach on local fisheries, depriving the people of their livelihoods and a critical source of protein. MSP could help by affording facilities for surveillance or providing information to serve both global and local purposes, including recognition on the part of participants that threats can move from one part of the maritime domain to another.

- *Congruity of strategic interests.* This consideration ranges from a general interest in uniting all the organized maritime protection services in the world in order to share information and assist one another as necessary, all the way down to organizing local police and enforcement actions. The broadest possible cooperation and sharing builds trust among countries and enables them to work together when needed.

A final note on the committee's understanding of why the United States is especially interested in MSP for the new era: It believes that 9/11 and the newly discovered need for homeland defense against Islamic extremists led the nation to take a greater interest in the maritime domain. American fears were compounded by the fear that such extremists might acquire and use nuclear weapons. This fear goes beyond the long-standing American concern for nonproliferation. The United States prefers that any such attacks take place as far away as possible. As the United States had long done, but more so as it moved onto the world scene after World War II and during the Cold War, it assumed responsibility for reducing conflict and unlawful activities around the world in the interest of general stability and rising prosperity for all peoples. Now with the spread of globalization and the attendant growth in mobility and communications, the United States finds a reason for reorienting its maritime outlook for both homeland defense and for the worldwide security of the maritime domain, especially against smaller and more scattered threats. If many other countries also perceive the need for such a reorientation, there may be a good rationale for expanding MSP.

Figure 1.2 arrays these considerations, from local through regional to global arrangements along one axis and from independent country efforts in their own sovereign waters to fully integrated efforts with other countries and international organizations along the other. For illustrative purposes a number of the organizations and international conventions that operate in the maritime domain to date are shown.

Understanding How to Implement an MSP Program

The committee understands that there are some general guidelines to be followed in implementing a program for MSP:

- Achieving MSP is a matter of communicating among the governments and maritime organizations of different countries, the international organizations

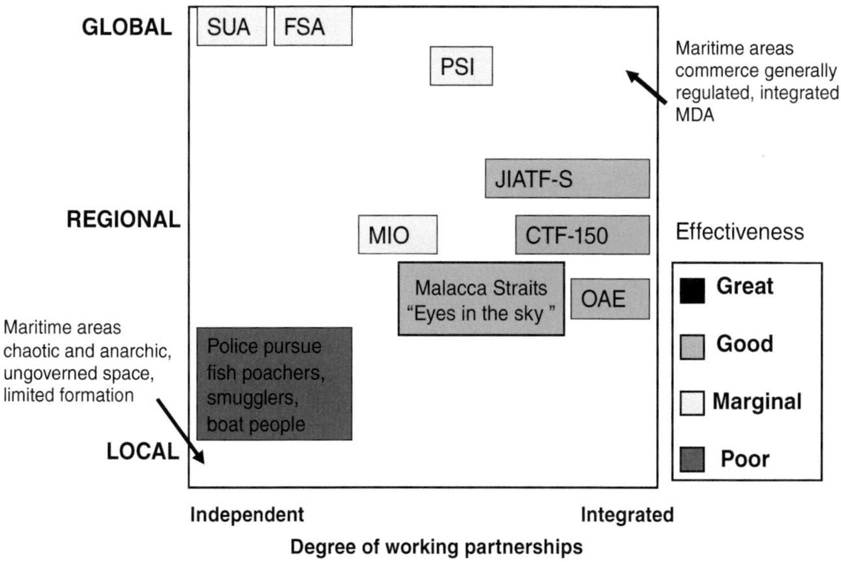

FIGURE 1.2 Maritime security partnerships. NOTE: A list of acronyms is supplied in Appendix G.

that have cognizance over maritime affairs, and the maritime companies whose protection is the intent of all this regulation and law enforcement.

- First of all, the U.S. government must organize itself for this effort. The U.S. Navy obviously proposed the 1,000-ship Navy, but it is not just a Navy effort, especially given the many maritime organizations (navies, coast guards, and so forth) that would be involved around the world and given the diplomatic efforts required to arrange the cooperation. On the U.S. side the effort must be a U.S. government interagency one, especially given the participation of the USCG.
- In short, the Navy enlists the Secretary of Defense, who in turns gets approval from the President, who then directs other U.S. departments and government agencies (State, USCG, Commerce, and others) to participate. The State Department, in turn, approaches other countries and international organizations, as appropriate, to enlist them. The countries instruct their various maritime organizations to consult with the U.S. organizations to decide on the most satisfactory arrangements and a course of execution. In addition, the State Department, along with other cognizant U.S. organizations, opens consultations with the relevant international organizations, like the IMO (the USCG would be the working contact here).
- The U.S. country teams, the U.S. combatant commanders (COCOMs), and probably the USCG will play significant roles. They will advise on the best way to approach countries. The COCOMs can use their theater security coopera-

tion (TSC) plans to provide exercises, training, and equipment to other countries to support them in policing the maritime domain. The USCG may be best positioned to deal with the variety of coastal and port maritime law enforcement organizations. In the long run, the U.S. Navy will be developing a cadre of foreign area officers (FAOs) who can serve on country teams in the security assistance organizations and liaise with local maritime organizations, especially those in the less-developed countries.

• Organizing and extending an MSP program is going to be an evolutionary, organic process—it will not be possible to design a complete architecture all at once from the beginning. However, the program can be based for the most part on existing cooperative arrangements. The goal of MSP should be to enable countries to act locally to solve their own problems, then to begin talks and work toward regional associations, and finally to tie the regional associations into a broader, globally networked, maritime-information-sharing cooperative. This process is illustrated in Figure 1.3.

• An informal model of organization, similar to that of the Proliferation Security Initiative (PSI), should generally be followed. While there might be

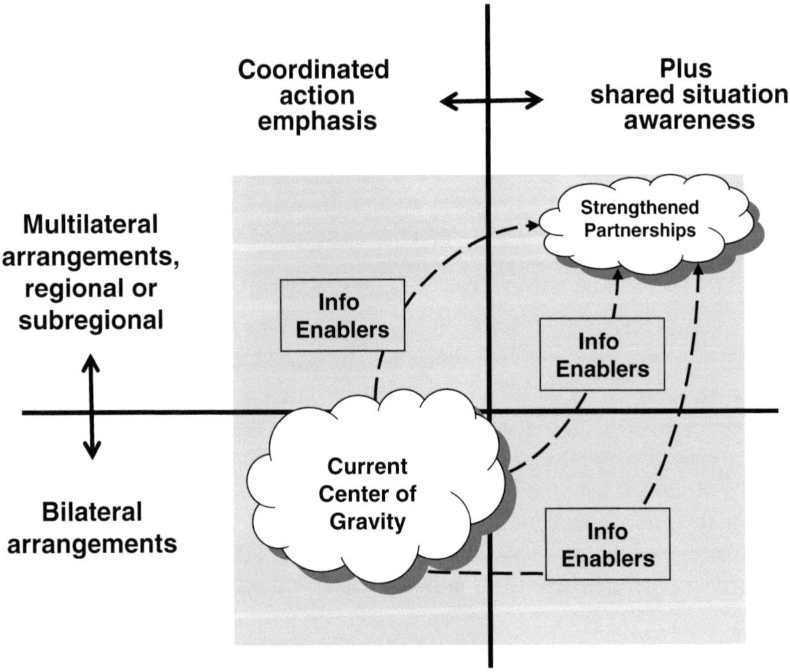

FIGURE 1.3 Current and emerging international maritime security partnerships.

mechanisms for coordinating surveillance, transferring information, and so on, there would be no central headquarters or staff beyond existing organizations such as the IMO. Rather, any organization that handles the information arrangements would be acting like a telephone exchange.

- Maximum advantage would be taken of existing arrangements, alliances, and governmental and international organizations. Several international models for MSP cooperation already exist, as shown in Figure 1.2. The COCOMs already have working relationships and programs under their TSC programs.

At the same time, as these MSP arrangements unfold, it is likely to become apparent that some of the international legal conventions, especially those that relate to boardings, whether in sovereign or international waters, will need to be clarified and extended in their authorities. This would take a more formal process—one, for instance, that accords with IMO procedures.

THE COMMITTEE'S APPROACH TO ADDRESSING THE TERMS OF REFERENCE

As outlined earlier in this chapter, as the study progressed, the committee refined its understanding of the four levels of cooperation for maritime security described in the terms of reference. It became evident to the committee that these four levels of an effort for maritime security are already in operation in various parts of the world. For example, at the first level, the U.S. Navy, the USCG, and merchant shipping are already cooperating on MDA initiatives, where much of the current activity entails the installation of automatic identification systems (AISs) on all vessels over 300 GT. At the second level of cooperation—U.S. naval and maritime assets with those of other countries with which it has treaty alliances or analogous arrangements—the Joint Task Force-150 Combined Enterprise Regional Information Exchange System (CENTRIXS) shares secret information among traditional NATO members and coalition maritime partners in Operation Enduring Freedom. At the third level of cooperation—U.S. naval and maritime assets with ad hoc coalitions—the Cooperating Nations Information Exchange System (CNIES) is being used by JIATF-S and 11 cooperating nations in South and Central America in efforts to suppress illicit maritime drug traffic. At the fourth level of cooperation—U.S. naval and maritime assets with others than those above that may now be friendly but could become hostile, for special purposes such as deterrence of piracy or other criminal activity—demonstration networks for the sharing of unclassified, commercially available AIS (and other) information with and among nontraditional partners are in progress in the Gulf of Guinea Initiative between U.S. naval forces in Europe.

In reviewing the existing and emerging international partnerships, it became clear that one size does not fit all in the matter of information sharing and enabling technical mechanisms for maritime security. Differences are traceable to:

- Differing levels of trust,
- The distinction between bilateral and multilateral arrangements,
- A focus on coordinated tactical-level action rather than information sharing, and
- Differing levels of technological maturity and sophistication.

Also, the center of gravity for current maritime partnerships resides in bilateral arrangements for the coordinated execution of tactical actions supporting common security interests (e.g., interdiction).

Thus, as the study progressed, the committee discovered that a different organizing principle was more appropriate to the complexity of the 1,000-ship Navy concept as it is being developed and implemented. The committee's approach was to add value to the ongoing efforts and respond to the spirit of the CNO's request while at the same time addressing the four levels of cooperative effort for maritime security and the four tasks (the four bullets) in the terms of reference. Furthermore, discussions with the Deputy Chief of Naval Operations for Information, Plans, and Strategy (N3/N5) at the committee's first meeting encouraged a broader approach to the study, one that would address the more important question of how to achieve MSP. This carried the effort beyond the question of how to support MSP technically and operationally and concentrated instead on how to attract the wide range of participants suggested in the terms of reference. With this in mind the committee used a somewhat more complicated set of bilateral and multilateral models of cooperation to address the tasks in the terms of reference.

ORGANIZATION AND CONTENT OF THIS REPORT

With the understandings, assumptions, and approach specified above, the committee organized its response to the terms of reference as follows.

Chapter 2 examines previously established models and other possible operational concepts for different levels of cooperation, including both the NATO and the Interpol models called for in the first bullet item of the terms of reference. It goes on to discuss the agreements, laws, and treaties for building partnerships; concludes that trust between partners is key to success in MSP; and provides recommendations for building and expanding the partnerships, both domestic and international, that are needed for successful implementation of MSP.

Chapter 3 addresses the second, third, and fourth bullet items of the terms of reference. It goes on to conclude that improved information systems and technologies and the associated exchange mechanisms for information sharing are key to achieving success in implementation of MSP and recommends such improvements.

Chapter 4 discusses the roles and responsibilities of the U.S. Navy and others, creates an implementation strategy for MSP, including force structure (second bullet in the task statement), and recommends mechanisms for improved governance of maritime security.

Appendixes A through G provide supplemental and study-process-related information.

2

Maritime Security: Cooperation Modes and Models

The maritime security partnerships (MSP) initiative seeks to develop cooperative arrangements between countries that allow them to share data among themselves to improve the situational awareness of activities off the shores or borders of those nations. States can then decide to act independently or cooperatively if they choose to address what they perceive as a threat to their security or the security of one or more of the other parties.

The premise of the MSP initiative is that by improving its awareness of what is happening in maritime areas that could be of interest to it, a state directly improves its security and would therefore be willing to share similar data with those countries it perceives to have congruent interests. Relationship building and information sharing during normal times may also mean that in time of crisis, the state will be able to call upon individuals or information to address an emerging problem. The ease and trust with which information or individuals can be accessed will be directly related to the success of their past relationship.

NEED FOR INTERNATIONAL LEGAL FRAMEWORK

The key to effective MSP is improved maritime domain awareness (MDA)[1] among participating states, which—along with agreements to take coordinated, mutually supportive tactical actions—will enable them to address, individually or

[1]The Department of Homeland Security's 2005 *National Plan to Achieve Maritime Domain Awareness* (Washington, D.C., October, p. 1) defines MDA as "the effective understanding of anything associated with the maritime domain that could impact the security, safety, economy, or environment of the United States."

collectively, what might be called "existing gaps in global maritime governance."[2] The existence of such maritime governance gaps has been made painfully evident by patterns of activities or events at sea that raise serious and legitimate concerns on the part of directly affected states and the international community at large: armed attacks on shipping; acts of piracy and terrorism; maritime trafficking of weapons, people, and drugs; marine environmental pollution; and illegal, unreported fishing. These phenomena are readily attributable to the sheer vastness of ocean spaces, the huge number of vessels involved, the lack of transparency that characterizes the maritime industry as a whole, and the comparatively limited resources that individual states can bring to bear on these problems. In the final analysis, they all point to inadequate information and inadequate resources, with the former pointing to MDA as an indispensable enabler of maritime security.

Efforts to bolster the acquisition, processing (analysis and fusion), and sharing/distribution of maritime information—in short MDA-related core activities—have significant implications for the international legal system. Given the broad range of conceivably relevant MDA-supportive measures, from off-shoring of security measures at one end to the nonconsensual boarding of a foreign flag vessel at the other, the drive to improve MDA is likely to affect the existing balance of power between flag states on the one hand and coastal and port states on the other. This balance has found expression in an elaborate set of rules that today are reflected principally in the United Nations Convention on Law of the Sea (UNCLOS) and customary international law and in some other maritime treaties.

Success in garnering wide international support for the idea of MSP—a critical precondition if it is to be effective—will depend on the proponents' ability to demonstrate convincingly that the common interest of all states is being served by MSP. Success will similarly require an approach for lobbying other states, international organizations, and civil society in general—in short, a judicious choice of implementation strategies and tools. What is less appreciated, however, is that the willingness of states and other actors to endorse and actively participate in MSP will also depend on whether they perceive the arrangement to be internationally legitimate. Indeed, concerns about legitimacy may turn out to be the stumbling block to the realization of the global maritime security network. For MSP to succeed, states must either come to see the project as compatible with existing international legal frameworks and rules or, conversely, understand that MSP proponents are willing to seek the adjustment of applicable legal rules, if necessary, to accommodate MSP within the international legal structure. The cautious attitude of several key states to signing up for the Proliferation Security Initiative (PSI)[3] and similar attitudes expressed by representatives of the Indian

[2]See Chapter 1.

[3]Thus far, several states whose support of MSP would be extremely important—for example, China, India, Indonesia, and Malaysia—have not joined PSI, and Russia's participation is conditional.

and Chilean navies as well as the Royal Navy in briefings to the committee on the topic of MSP, show that without a solid international legal grounding MSP is unlikely to reach its full potential. More details of the international legal framework for MSP can be found in Appendix C.

MODELS FOR MARITIME SECURITY PARTNERSHIPS

The United States and most other countries participate in numerous information-sharing arrangements with varying degrees of trust and kinds of information. Traditional military missions, particularly during the Cold War, relied heavily on institutionalized modes of cooperation under formal treaties against a background of extensive operational activities and persistent arrangements. For the United States, the North Atlantic Treaty Organization (NATO) is still the most successful model for cooperation among states. The reach of NATO in terms of cooperation, information sharing, and equipment standardization goes far beyond the formalities of the fairly standard NATO treaty. It has been enriched by more than 50 years of experience, negotiation, and trust building. Even in times of discord and disagreement, NATO maritime forces train and exercise together and, especially in the last decade, have engaged in the full range of joint operations, with assignments in the Persian Gulf, Bosnia, the Arabian Sea, along the East African coast, and now in support of Afghanistan, although not always as formal NATO missions. The regular sharing of intelligence and data from surveillance and reconnaissance surveys is the stuff of daily life for most navies, even those of the new members in central and eastern Europe.

A less familiar model is the innovative but limited partnership developed by the United States and the Soviet Union/Russia in the heyday of strategic arms control. In narrow areas, information of strategic significance was defined, exchanged, and even jointly developed. An elaborate vocabulary of signals, formal and informal, emerged, along with specialized protocols and mechanisms (such as the Washington-Moscow hotline) for risk avoidance or crisis dampening. Regular meetings and continuing negotiations raised the level of information exchange and even led to a sharing of terms of the trade in negotiation and reconnaissance. "Trust but verify" became not only a watchword but also a standard for the type of information sharing that took place between partner states that were never quite friends but not strictly adversaries.

The Cold War models with traditional partners do not adapt easily to the requirements of cooperation with nontraditional partner states to address nontraditional maritime security threats such as terrorism, economic crimes, piracy, civil turbulence, or failing state governance. But the United States and a number of other countries have had a wealth of experience over the last two decades that suggests forms and procedures to be followed in developing and securing MSP.

PREREQUISITES FOR MARITIME SECURITY PARTNERSHIPS

It is this committee's observation that there is no one single model that must or should be used in forming MSP. Many potential nontraditional partners for maritime security facing new challenges do not need, nor could they operate, elaborate programs or mechanisms that conform to present NATO standards, for example. Rather, they need purpose-driven programs that help with maritime situational awareness and capacity building. The models that now exist in the maritime and other domains, such as the programs developed from 1994 to the present under NATO's Partnership for Peace (PFP), provide a rich source of experience and information. Fundamentally new models are emerging as well, such as the robust new Gulf of Guinea Initiative under Naval Forces-Europe and the multinational work in Joint Task Force-150 operating in the Arabian Sea.

The technology needed to establish networks for information exchange with nontraditional partners is already widely available or relatively easily adaptable to existing equipment; no elaborate new system development seems required. The main constraints are (1) rather outdated domestic legislation on foreign information sharing and export control procedures in the United States; (2) inadequate domestic information sharing and program planning; and (3) the low priority accorded to the primarily nontraditional challenges of this century.

The sum of these experiences persuaded the committee that three critical elements are needed to achieve local, regional, and global success in establishing new MSP or improving existing ones:

1. A cadre of trained, proactive specialists, military and civilian, who are able to operate linguistically and culturally in the region or in the U.S.-based planning and coordinating functions—as, for example, in the reestablished Navy foreign area officer (FAO) program or the FAO programs that already exist in the Army, the Air Force, and the Marine Corps (see below).
2. Secure, persistent funding that is adequate in the immediate future to support particular opportunities. For example, to secure the transition away from Soviet-era military training and equipping models in central and eastern Europe in the early 1990s, the United States increased funds under the PFP process and labeled its action the "Warsaw Initiative," which then for more than a decade and a half supported expanded military-to-military exchanges and exercises among new NATO members and PFP candidates.
3. A robust coordinating authority, particularly at the highest levels of the U.S. government, that can arrange appropriate governance at all levels (see Chapter 4). It could bring disparate program elements in from across the different agencies and ensure a proactive, coordinated effort to overcome local challenges while also expanding planning and integrating domestic and international priorities (as, for example, the Container Security Initiative).

Several elements influence the form of prospective MSP agreements and the most appropriate time to get such cooperation agreements in place, including the following:

- *Level of organizational coordination/contact.* Is it high, medium, or low in the country's hierarchy of information organizations?
- *Commitment to consult.* Will the partners exchange information only in a defined situation? Will they do it sometimes or always?
- *Length of agreement or cooperation.* Is it a one-time arrangement for a specified time or a permanent arrangement that needs to be formally canceled?
- *Scope.* Are the partnerships local, regional, or global?
- *Military status.* Are the participants military or nonmilitary or both?
- *Primary area of activity.* Is the purpose mainly traditional defense, law enforcement, humanitarian, or commercial?

All these factors will impact the prospective scope and level of agreement.

THE RANGE OF PRESENT MSP RELATIONSHIPS

Figure 2.1 shows the range of agreements in which the United States and key maritime states currently participate. Most maritime partnerships are, as in the past, bilateral agreements, although many are nested within multilateral treaty frameworks to which the states already subscribe (see Appendix C). They are arrangements between two states that agree to provide each other information (a two-way exchange of information) for a specified purpose dictated in the agreement or treaty. The purpose can involve many different kinds of interaction, from a simple exchange of information or data to the other extreme, whereby one country would allow another country and its assets (say, vessels with embarked personnel) to enforce laws within its coastal waters.

Bilateral Relationships

Table 2.1 lists what the committee considers some of the more interesting contemporary examples of bilateral relationships relevant to maritime security and characterizes them according to the factors just mentioned. These range from those that are at a fairly basic level of information sharing and interaction to those that involve a wide range of tactical operations and cooperation—as enhanced cooperation and intensive efforts to develop common or converging bases for joint action, perhaps encompassing intelligence, surveillance, or reconnaissance at the very highest ends.

The process of developing relationships often starts with military-to-military contacts, which then lead to personal exchanges and contacts or a dialogue on a specific area of interest. In the Navy, it ranges from contacts between the Chief of

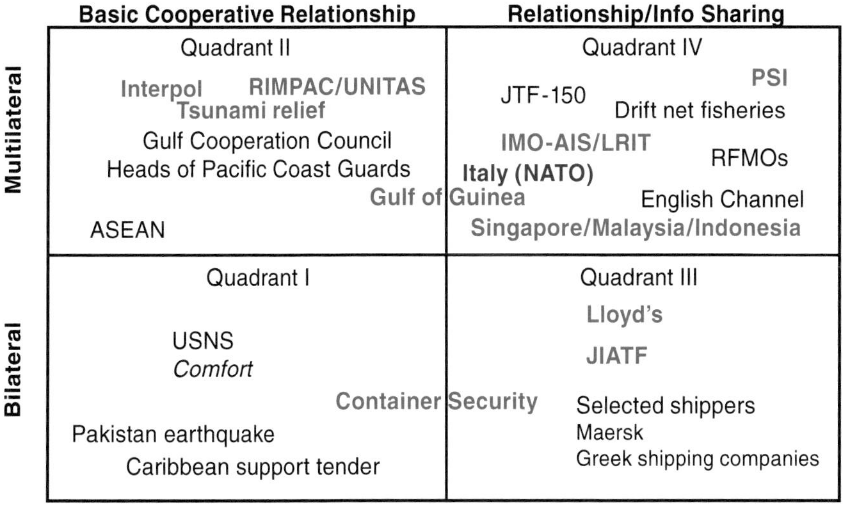

FIGURE 2.1 Types of agreement. NOTE: Interpol, International Criminal Police Organization; RIMPAC/UNITAS, Rim of the Pacific/Annual U.S.–South American Allied Exercise; ASEAN, Association of Southeast Asian Nations; JTF, Joint Task Force; PSI, Proliferation Security Initiative; IMO-AIS/LRIT, International Maritime Organization–Automatic Identification System/Long-Range Identification and Tracking; RFMO, Regional Fisheries Management Organization; USNS, U.S. naval ship; JIATF, Joint Interagency Task Force.

Naval Operations (CNO) or the fleet commander and the foreign Navy or Marine Corps counterparts; other relationships develop through the networks established during joint exercises, training, and port visits. For the USCG, it might be the Pacific Area Commander reaching out to his international Coast Guard counterparts on how to secure international trade lanes or it could begin like the work of the USCG with its Chinese counterparts in search and rescue exercises, securing trade lanes, and cooperating on maritime security and safety.

Such contacts can result in the signing of bilateral or even multilateral agreements. The United States for most of its history but particularly of late prefers to enter into bilateral rather than multilateral agreements. The basic reason for this approach is that an arrangement reached by multilateral consensus often ends up too watered down to mean much. On the other hand, multiple country-to-country (bilateral) agreements may end up raising everyone's boat in terms of the actions desired by an even larger group of potential partners.

It would be useful here to review the many bilateral counterdrug agreements the United States has concluded with countries of the Caribbean basin under the Joint Interagency Task Force-South (JIATF-S) (see Appendix D for further details of this and other programs). The discussions, initiated by the U.S. State Department at the request of the USCG, all begin with the United States presenting a

TABLE 2.1 Bilateral Relationships

	Level of Organization	Commitment to Consultation	Length of Cooperation	Scope	Military or Nonmilitary	Law Enforcement, Defense, Humanitarian, Commercial
USNS *Comfort*	L	S	O	G	M	H
Container Security Initiative	M	A	P	G	NM	LE
Pakistan earthquake relief	L	S	O	L	M	H
Caribbean support tender	L	S	S	R	M	LE
Lloyd's of London	H	A	P	G	NM	C
Joint Interagency Task Force-South	H	A	P	R	NM	LE
Maersk	M	S	S	G	NM	C
Greek shipping lines	H	A	P	G	NM	C

NOTE: H/M/L, high, medium, low; O/S/A, onetime, sometimes, always; O/S/P, onetime, specific, permanent; L/R/G, local, regional, global; M/NM, military, nonmilitary; LE/DEF/H/C, law enforcement, defense, humanitarian, commercial; USNS, U.S. Naval Ship (civilian manned).

model eight-part bilateral agreement that lists the full suite of joint purposes for which information may be shared. Washington's intention is for all agreements to look the same. In reality, very few are the same, because each country has different motivations and sovereignty concerns, at least in the initial phases. At some point, if bilateral accords are signed, as they were for the counterdrug efforts in the Caribbean, the desired end result is more nearly achieved with a multilateral agreement, which tries to accommodate all the divergent views.

The Caribbean counterdrug efforts that are directed by JIATF-S in Key West, Florida, operate multilaterally yet take into account all the separate bilateral agreements. This approach engenders trust and cooperation while showing an appreciation for the uniqueness and domestic politics of each country. There is now a wealth of examples in which implementation was specifically assigned to the state that had not only the capabilities for action but also the right rules of engagement as set by its national authorities. Such activities, observed and acted on over a significant period of time, have increased cooperation, trust building, and information-sharing activities.

Multilateral Relationships

The second form of agreement is a multilateral one, which is a single agreement signed by multiple nations and involving mutual commitments among all the participants (see Table 2.2).

Like bilateral agreements, multilateral agreements can entail a simple exchange of information or the use of force in a third nation's territorial waters in defined situations. They may also be more intense, more specific forms of earlier, broader multilateral agreements that set looser standards for action and cooperation.

International or global organizations that operate by treaty or convention represent a special, formal variant of multilateral agreements and most often impose not only a formal, global level of organization but also obligations, and they may convey rights under international law to all states and nongovernmental entities that participate. Examples of a range in duration and in formality of an organization are the multilateral agreements that set up Interpol, PSI, Joint Task Force (JTF)-150, and the International Maritime Organization (IMO), particularly its Automatic Identification System (AIS) and Long-Range Identification and Tracking (LRIT) system (see Figure 2.1). The last two agreements illustrate another critical function of international and multilateral organizations, the creation of universal standards that all signatories are pledged to meet. In this case, to ensure maritime safety and security throughout the maritime commons, IMO signatories have agreed to accept the relevant International Convention for the Safety of Life at Sea (SOLAS) amendments establishing standards for the collection of identification data for large ships (300 GT and larger) wherever they are, accessible through an agreed mechanism and in standard format. The IMO is also

TABLE 2.2 Multilateral Relationships

	Level of Organization	Commitment to Consultation	Length of Cooperation	Scope	Military or Nonmilitary	Law Enforcement, Defense, Humanitarian, Commercial
Interpol	H	A	P	G	NM	LE
RIMPAC/UNITAS	H	S	S	R	M	DEF
Tsunami relief	L	O	O	R	NM	H
Gulf Cooperation Council	M	S	P	R	M	DEF
Gulf of Guinea Initiative	M/H	S	S	R	M	LE
Heads of Pacific Coast Guards	M	S	S	R	NM	LE
Proliferation Security Initiative	M	S	P	G	NM	LE
ASEAN	M	S	P	R	NM	LE
Joint Task Force-150	H	A	S	R	M	DEF
Drift net fisheries	M	S	P	R	NM	LE
IMO-AIS/LRIT	H	A	P	G	NM	LE
RFMO	L	S	S	R	NM	LE
Italy (NATO)	M	A	S	R	M	LE
English Channel	H	A	P	R	NM	LE
Singapore/Malaysia/Indonesia	H	A	P	R	M	LE

NOTE: H/M/L, high, medium, low; O/S/A, onetime, sometimes, always; O/S/P, onetime, specific, permanent; L/R/G, local, regional, global; M/NM, military, nonmilitary; LE/DEF/H/C, law enforcement/defense/humanitarian/commercial; RIMPAC/UNITAS, Rim of the Pacific/Annual U.S.-South American Allied Exercise; ASEAN, Association of Southeast Asian Nations; IMO, International Maritime Organization; AIS/LRIT, Automatic Identification System/Long-Range Identification and Tracking; RFMO, Regional Fisheries Management Organization; NATO, North Atlantic Treaty Organization.

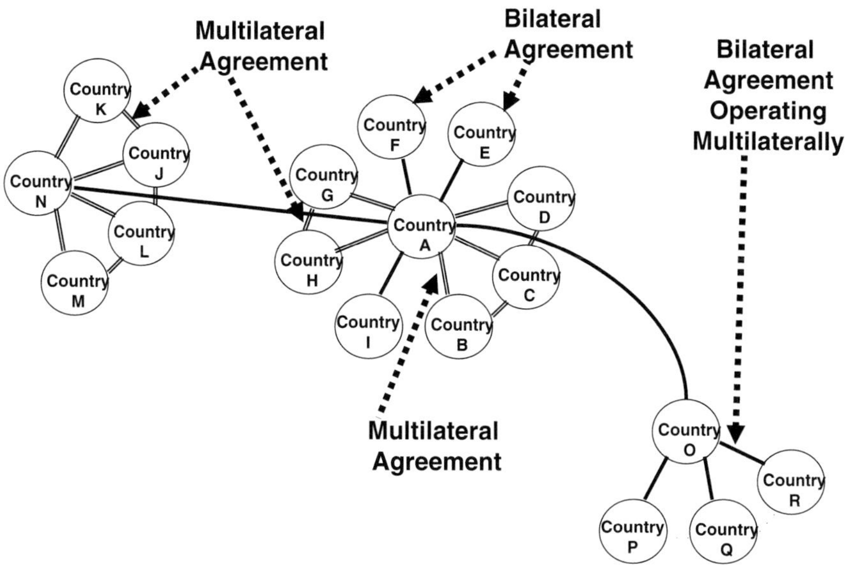

FIGURE 2.2 Agreement types.

in the process of specifying a global system for the storage of the data collected and disseminated to states and interested parties according to specific rules and protocols under international law.

Multilateral organizations might also spark the establishment of informal but widely accepted norms for behavior or standards for action. While such norms may not be accepted by all states and hardly can be said to have been established by formal agreement, bilaterally or multilaterally, they raise expectations about what should be done, expectations on the part of ordinary people or the media, and critical public and private actors if not always governments.[4]

Figure 2.2 shows that it is possible for bilateral and multilateral agreements to be in place with various countries at a given time. The relationships shown in color represent multilateral agreements, while bilateral agreements are shown by the solid black lines between the countries. Country N is depicted with a single multilateral agreement, and Country O is shown with multiple bilat-

[4]This accretion of legitimacy is thought by some to be the first step in the creation of what international relations specialists consider an informal regime, still not lawlike rules but a cluster of ideals and behavioral metrics. It parallels the way in which international prohibitions on the slave trade began or the growing expectation throughout much of Europe that individuals taken into police custody should be read their Miranda rights even though these rights formally apply only to U.S. citizens in U.S. jurisdictions. Analysts speculate that this reflects the constant invocation of this process in television programs and movies screened abroad.

eral agreements. Country A and Country L have both bilateral and multilateral agreements.

The challenge of the multilateral agreement is getting all parties to agree on the similar terms and conditions—often an arduous process. The bilateral agreement is preferred because it expedites the approval process between the parties and puts in place a mechanism for beginning to cooperate. Indeed, for some purposes, throughout its history and certainly in the last 8 years, the United States has preferred the use of informal agreements. The important thing is to address the challenge and not get bogged down in a prolonged bureaucratic negotiation.

The MSP initiative is as much about the network and services that maritime security agreements provide as it is about the trust and cooperation that are built through the networks. The network permits sharing of information under rules agreed to by the signatories. In the service-oriented architecture model, the parties agree to post MDA information based on their observations. Other countries can be authorized to receive this information and post their own data. As in NATO, these access rules need not be symmetrical or identical at the outset, although over time the arrangements tend to converge (as they did in the JIATF-S). As this process evolves and countries become comfortable with the interactions and data sharing, they begin to build trust and broaden or deepen cooperation and start to benefit from the mutual activities.

The Geographic Reach of MSP

Many, including some recent participants, argue that the difficulty and delays in getting these agreements approved grows directly with the number of nations involved. Agreements and functional cooperation can be local, regional, or global, depending on the scale of the challenge and complexity of the approval process for the agreement. Local agreements in a small geographical area—say, agreements about illegal fishing or piracy or the mutual right to arrest citizens of either country who break safety or environmental laws in the territorial waters of either state—typically involve a law enforcement arrangement, usually within the maritime environments of two or three nation-states. Regional agreements involve neighboring states that come together to address a common problem. The Association of Southeast Asian Nations, a relatively informal regional state-to-state network with growing cooperation on security, is a good example of this type of agreement. Finally, global agreements can involve nation partners, nongovernmental associations (e.g., shippers' associations), or commercial entities (e.g., Lloyd's) in addressing global issues. All these types of partners participated, for example, in persuading the IMO to accept the AIS standards for tracking ships larger than 300 GT.

Regional approaches seem to work best when interests are congruent and the stakes are clear or when legacy practices or habits of cooperation can be extended to support present agreements. Global effects are often desired and can be articulated when one of the affected states is much stronger than the others. But for most maritime security purposes, the committee believes that all issues are local.

In some cases, a relationship that is formed to address a specific immediate issue starts out as an ad hoc relationship, but if it succeeds it may become a formal program. The U.S. government's response to the tsunami in the Indian Ocean is an example of an urgent, ad hoc relationship formed to respond to the devastation. A task force was organized and moved to the area, where it engaged the affected governments—India, Indonesia, and Thailand—and provided whatever relief it could. The end result was very positive, and the view of the United States was enhanced by the manner in which aid was provided. In Indonesia, for instance, only 30 percent of the population viewed the U.S. government favorably before the assistance was rendered. Afterward, a favorable impression was shared by 70 percent. Also, as a direct consequence, military-to-military contacts, which had been suspended, were resurrected and still continue. The United States had a similar experience when it assisted Pakistan after the earthquake in 2005.

To sustain the favorable opinion, the U.S. Pacific Command (PACOM) has worked aggressively to further these nontraditional contacts involving maritime personnel. A 2007 event was the deployment under specific bilateral agreements of the USNS *Comfort* hospital ship to the region to provide humanitarian assistance. As a result, the U.S. Navy and DOD are poised to respond to international natural disasters around the world as a means to change impressions and develop enduring positive relationships.

Special Cases

Sometimes offers of cooperation have unintended consequences. The increased incidence of piracy in the Malacca Strait and its impact on international shipping flows and insurance led ADM Fargo, USN, PACOM commander, to propose that the U.S. Navy might help to patrol in the area. The governments of Singapore, Indonesia, and Malaysia agreed to undertake patrols and share information in order to secure the area and said that U.S. Navy patrols would not be needed or welcome. The trilateral Malaysia, Singapore, and Indonesia (MALSINDO) scheme has been successful and has added a fourth partner, Thailand. It now shares information with PACOM and has multilateral ties with eight other partner states under the Regional Cooperation Agreement on Combating Piracy and Armed Robbery against Ships in Asia (ReCAAP, signed in 2006).

Some programs of cooperation do not succeed directly but set the stage for other programs later on, when political conditions have become more auspicious. The Navy's Fleet Station concept, now being planned for the Gulf of Guinea, is rooted in an earlier cooperative approach in the Caribbean, the Caribbean support tender (CST). Under bilateral agreements and through the State Department, the USCG provided regional states with decommissioned vessels—specifically, 82-ft patrol boats and buoy tenders. It soon became apparent that without assistance, these vessels would not be maintained and would not have the desired force multiplier effect. The USCG's CST then became a circuit rider throughout the region, providing training and maintenance.

A unique aspect of this program was the personal relationships that developed among the international crew of approximately 50. The captain and a small cadre of officers and enlisted men were from the USCG, and the remaining crew comprised member country officers and enlisted men who sailed on the vessel for a year, returned to their own countries, and assumed positions of leadership. There was much competition to become assigned to this vessel.

Cooperative relationships build on the power of examples and available models. In the buildup to and during the initial phases of Operation Iraqi Freedom in 2002, the coalition partners saw the need to protect U.S. and coalition assets transiting the Strait of Gibraltar and set up a task force to do this. It patrolled under Operation Active Endeavor and is still functioning today. In operations, it is essentially the old STANAVFORMED (Standing Naval Force Mediterranean) and STANAVFORLANT (Standing Naval Force Atlantic) combined, taking on a specific mission rather than just a training exercise. Other countries have joined its patrols, including Russia. It has essentially operated in the western Mediterranean and was designed to intercept ships carrying materials for weapons of mass destruction and has queried thousands of ships for this purpose. However, the actual boardings have been to disrupt the north-south flow of illicit goods and people into southern Europe.

FINDINGS AND RECOMMENDATIONS

The Gulf of Guinea Initiative is probably the most interesting example of a combatant commander (COCOM)-driven program to broaden and thicken maritime security partnerships in the face of turbulent regional challenges (see Appendix D for specific details).

One example (without U.S. participation) of making a relationship more effective is the multilateral organization that has developed in the Strait of Malacca and the Strait of Singapore. Maritime security for Singapore is all about national survival. As a consequence, Singapore has become proactive in policing the straits and other local waterways. Patrols with Indonesia were first initiated in 1992. In July 2004, the trilateral organization MALSINDO started the Malacca Strait Security Initiative (MSSI) patrols in the Strait of Malacca. In September 2005, the MSSI organization became quadrilateral when Thailand joined and the "Eyes in the Sky" program was begun with sensor-laden aircraft overflying the Strait of Malacca. Piracy has been significantly reduced. ReCAAP, an agreement that includes Cambodia, Japan, Laos, Singapore, Thailand, Philippines, Myanmar, and South Korea, has been in force since September 2006.

Out of these efforts, a supporting information infrastructure has emerged: the Vessel Traffic Information System, which receives inputs from the closed-circuit television surveillance system; AIS transponders; and the Singapore Port Traffic Management System. Since January 1, 2007, all licensed powered harbor and pleasure craft are required to have the Harbour Craft Transponder System (HARTS), which feeds into the Port Operations Control Center. The Regional Maritime Information Exchange (ReMIX) is targeted at the Western Pacific Naval Symposium (WPNS) operations community. Information is exchanged on robberies at sea, piracy incidents, missing or hijacked ships, vessels in distress, and other maritime events. ReMIX is a Web-based platform, accessible by password. Once logged in, navies are free to upload and download information as they need. Information is shared with the Combined Enterprise Regional Information Exchange System (CENTRIXS), made up of nearly 30 countries throughout the world, including the United States.

The Singapore Navy's Access System is a portable command and control (C2) system that allows sharing a sea situation picture among the various nations. It uses the commercial-satellite-based Global Positioning System and proprietary C2 software for the automatic tracking of targets. It has a chat-and-file transfer facility for target management. The Singapore data-sharing structure, in partnership with PACOM, is shown in Figure 2.3.

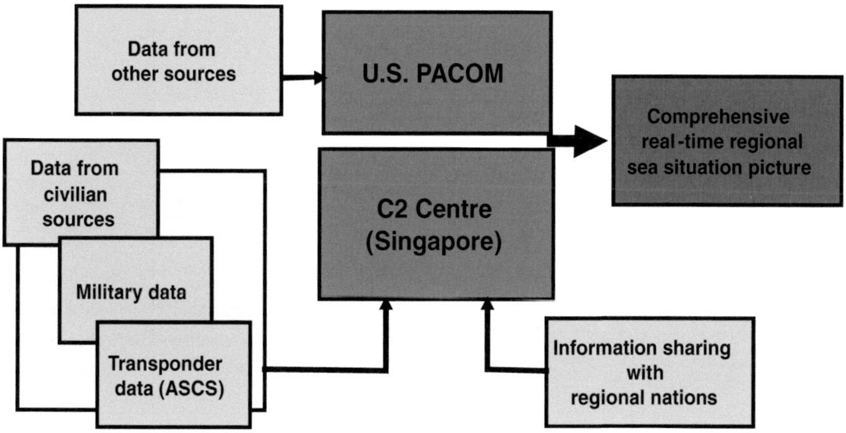

FIGURE 2.3 Singapore's organization for information sharing. SOURCE: COL James Soon, Republic of Singapore Navy, Head, Defence Technology Office, Embassy of Singapore, "The 1,000-Ship Navy: A Perspective from Singapore," presentation to the committee, Washington, D.C., March 14, 2007. NOTE: U.S. PACOM, U.S. Pacific Command; C2, command and control; ASCS, Acoustic Sediment Classification System.

Finding: Most information-sharing relationships start out as an individual bilateral agreement between the United States and one other country. The greatest gains in the intermediate term come from expanding bilateral relationships and agreements. In many cases, the base on which to build will be military-to-military relationships that can be expanded to include other groups—military and civilian, government and nongovernment—that are important to the maritime security task.

Recommendation 1: The Chief of Naval Operations, working with the combatant commanders, the Commandant of the Coast Guard, and the Commandant of the Marine Corps[5] should commit to transforming bilateral relationships into broader, more substantiative and inclusive maritime security partnerships by some or all of the following means:

- Forward presence;
- Increased language and cultural awareness;
- Expeditionary training teams;
- Ongoing analysis of gaps in capacity with plans for follow-up capacity-building steps;
- Tools and resources appropriate for the particular geography of an area—for example, shallow draft vessels such as the HSV-2 *Swift* rather than larger and deeper draft combat vessels;
- Maritime domain awareness—information-sharing systems that will eventually be expandable to include both unclassified and classified information; and
- Funding for Phase Zero.[6]

Two quite different examples show the way in which national legislation can be put to use:

- *The Singapore example.* In 2003, the Singapore parliament adopted legislation designed to keep out of terrorist hands materials that could be used for making WMD devices. The Strategic Goods (Control) Act gives the government more legal muscle to track strategic goods—about 600 controlled items, including munitions and materials with civilian and military uses. Singapore is the first

[5]The identification of specific officers and offices in the government with specific recommended actions is intended to reflect those most closely aligned in terms of the existing structures of organizational responsibilities.

[6]The traditional four phases of a military campaign identified in joint publications are deter/engage, seize initiative, decisive operations, and transition. Phase Zero encompasses all activities before the beginning of Phase I—that is, everything that can be done to prevent conflicts from developing in the first place.

Southeast Asian country to have laws aimed at controlling the movement of such goods.

- *The Proliferation Security Initiative example.* PSI, established in 2003 in response to fears about the spread of WMD, represents a mixed form of intensive multilateral and bilateral information sharing. According to the Bush administration, it is not an organization or a treaty framework but an "activity." PSI is a set of interlocking bilateral and multilateral agreements among a group of almost 20 core supporters, associated with a set of declarations of support, many done secretly, by a larger group of more than 80 states.[7] The core supporters have subscribed to a Statement of PSI Principles, which includes a commitment to improve constraints at borders, ports, in the air, on land, and at sea by exploiting existing and new national legislation, as well as a commitment to consult and an implied willingness to take action if there is credible evidence of incidents in their sovereign territories. There have been a number of largely unpublicized actions under PSI, perhaps as many as 30 interdictions, including a few boardings at sea, in the past 4 years. A number of key countries (India, China, Indonesia, and South Korea) remain outside PSI, while Russia supports only the general concept of PSI activities.

PSI has been reinforced by a number of bilateral arrangements on specific issues. The United States, for example, has concluded agreements with seven flag states supporting and specifying the conditions for a U.S. right to board after a formal request has been made to board ships on the high seas suspected of carrying components for WMD. The ships of these seven states taken together with those of the PSI core states themselves account for more than 70 percent of the world's commercial fleets. PSI is also in congruence with (although deliberately not referenced by) two broad United Nations Security Council resolutions, 1540 (against nuclear terrorism and proliferation) and 1718 (action against nuclear developments in North Korea). It also will gain status if support grows for relevant amendments to the IMO Convention for the Suppression of Unlawful Acts Against the Safety of Maritime Navigation.

Finding: With each of the participants of the existing maritime security partnerships, there appear to be sufficient national and international legal frameworks to support the maintenance and the extension of maritime security initiatives.

Exercises and exchanges are fundamental vehicles of trust building that lead to nation-to-nation cooperation. Information sharing can be facilitated through combatant commander (COCOM) maritime operations centers or headquarters to develop awareness and to develop relationships with partner nations. Training for cooperation lends itself readily to gaming as an effective vehicle. Face-to-face

---

[7]See the list of PSI participants at the Web site of the Bureau of International Security and Non-Proliferation, available at <http://www.state.gov/t/isn/c19310.htm>. Accessed on September 25, 2008.

gaming with foreign partners will address the issues of cooperation before being forced to play in real time.

The instruments of operational cooperation range from equipment and systems to training of both U.S. and partner nation personnel in the COCOM's area of responsibility. Clearly, having the equipment and software systems both to interface with an information-sharing database and/or to feed the database is critical for all partners. Integral to this (as noted above) is the development of data standards for sharing.

Specific intensive course material and tactical gaming experience will have to become part of the curricula at all the Navy's professional schools—the Naval Academy, the Naval Postgraduate School, and the Naval War College, paralleled by similar actions at the USCG and Marine Corps professional schools. Emphasis should be placed on the opportunities and instruments that exist to develop and implement these partnerships and on the interagency opportunities and competencies. Consistent with the 2006 and 2007 decisions of the CNO on language and cultural enrichment education initiatives, these opportunities include those military officers now assigned to specialize in specific regions and languages.[8]

The core curriculum will aim to build a network of experts across the federal agencies and across the public-commercial divide who know and trust one another and who will have expectations about joint programs and cooperation. It will legitimize maritime security as a professional specialty, a military occupation specialty (MOS) that will give new prominence to the MSP concept.

While the relevant agencies are present to some degree within the military educational system, they are not always present in large numbers. MSP training will have to draw not only on the talents that exist at the Department of State but also on those that exist at the Departments of Commerce, Justice (specifically the Drug Enforcement Agency (DEA)), Transportation, Treasury, and Homeland Security. It will also need the strengths of the Navy's newly reestablished foreign area officer (FAO) program, parallel efforts within the Marine Corps, and a specialized program that the USCG should establish. An effort should be made to increase the number of those attending from each agency to between three and five per maritime professional school and to designate those with special skills as maritime security partnership scholars. Recognition for individuals who attend and go on to a successful career will be a sure indicator of the long-lasting relevance of this activity.

A smaller number of emerging civilian and military leaders should be selected to take part in a shorter training course specifically designed to foster networks and develop capacity across the interagency core involved in MSP. Lasting 3 to 6 months, with downstream refresher courses available onsite or electronically,

[8]The National Research Council's Naval Studies Board has just conducted a study of the manpower and personnel needs for a transformed naval force (see National Research Council, 2008, *Manpower and Personnel Needs for a Transformed Naval Force,* The National Academies Press, Washington, D.C.).

the curriculum should be designed for managers and implementers of maritime security partnerships, coming primarily from within the government but also from shippers and other relevant commercial companies.

Finding: The continued training of U.S. and partner nation personnel in a maritime security partnership is critical to long-term success and to building the relationships and trust that eventually result in the establishment of maritime security partnerships with as many countries as possible.

MSP requires the ongoing development of a cadre of military and civilian personnel to widen the scope of cooperation both within and external to the military. This will require training in maritime cooperation and an appreciation for the relevant competencies across the broader government and private sectors. Such appreciation for the capabilities of other agencies needs to become a core leadership quality, creating a diversified atmosphere that results in a multiplier effect in these nontraditional areas of military concern.

Recommendation 2: To educate and train U.S. and partner nation personnel so that they can support and extend maritime security partnerships, the Chief of Naval Operations should:

- With the active support of the leaders of the Marine Corps and the Coast Guard, ask the combatant commanders to support and extend maritime security partnerships through continued and even expanded formal educational and bilateral/multilateral training exercises for these personnel;
- Require that maritime security training become a significant part of the core curriculum at every level of professional education for maritime service;
- Adopt as a critical long-term goal the broadening of participation in maritime professional education to ensure representation from all of the relevant U.S. civilian and military agencies;
- Cooperate with the Secretary of the Navy and join in the present Coast Guard plan under the Department of Homeland Security to design and fund an institute of maritime studies that would encompass specialized studies in maritime security within the framework of an existing university program.

Critical for the longer-term ability of the CNO to implement MSP will be the establishment within the maritime services of a clear professional career track for officers and civilian officials with wide-reaching international expertise and experience. Appropriate models are the FAO programs of the Army and, to a lesser extent, the Air Force and the Marine Corps.

FAOs are individuals who select this as a military specialization early in their careers and train intensively in the cultural and linguistic skills needed for particular regions (e.g., East Asia) and/or functions (e.g., arms control monitoring and implementation). Later assignments, which may last longer (e.g., 4 to 6 years) than a normal tour of duty, may include assignment in an embassy, at a regional command, or as an in-country advisor to partner militaries. Most Army officers in these specialties accept as a consequence somewhat diminished career prospects (e.g., fewer opportunities for promotion to general officer), but a number have indeed gone on to flag rank by combining their capabilities.

Active CNO support for these programs will have a number of advantages, even though it may take 5 to 7 years to grow an initial group of Navy FAOs. These advantages would include not only career stability and enhancement and prospects of promotion but also official recognition of the value of their specialization and their particular contribution to the long-term maritime security of the United States and its partners. This could be particularly true for the Navy FAO specialization, which was introduced twice but failed to find a niche in the Navy's professional structure like that which it enjoys within the Army structure.

Until a maritime FAO cadre can be trained (this committee estimates that it will take at least 5 years), the CNO's mandate will have to rely on a "purple" or joint manpower approach (i.e., a resort to resources from the long-established Army FAO program). It might also draw on enlisted personnel with appropriate linguistic backgrounds, identified under the CNO's 2006 and 2007 directives on the identification of all Navy personnel with special language abilities and cultural awareness. Gaps could also be filled by civilian employees or contract personnel, who could provide the needed services at regional commands as well as at home.

Finding: There appears to be a shortage of qualified FAOs within the U.S. naval services. Such FAOs could provide invaluable aid in developing the capabilities of regional maritime security forces that would allow them to move their countries toward participation in regional and, later, global maritime information sharing.

Recommendation 3: The Chief of Naval Operations should mandate the expansion of a robust foreign area officer (FAO) program within the Navy to meet the needs of staffing and expanding maritime security partnerships. In addition, the Commandant of the Coast Guard should establish an FAO program and the Commandant of the Marine Corps should expand its present limited FAO program for the development of bilateral and multilateral relationships.

The law enforcement authority and legal skills that would be needed to carry out countersmuggling and counterterrorist activities in coastal waters do not usually exist aboard naval vessels. Naval vessels engaged in counter-drug-smuggling missions carry USCG law enforcement detachments (LEDETs) that actually board intercepted vessels that are suspected of smuggling drugs and, if needed, arrest their crews. Using personnel from the Naval Criminal Investigative Service (NCIS) or other law enforcement personnel could be equally effective, but additional training and equipment might be needed to gain ship boarding capabilities as well as to clarify the legal authorization.

If a LEDET is to be carried aboard a deployed naval vessel, that vessel in effect carries the full spectrum of U.S. maritime law enforcement and DOD authority. The present Maritime Operational Threat Response process can be used to determine under which authority an action is to take place. This additional onboard capability would give the U.S. government and, by extension, the COCOMs full-spectrum response capability. The USCG would need additional personnel and resources to carry out this additional tasking (these numbers might be available). At a later stage, the goal is to expand the onboard representation in law enforcement detachments to include selected interagency personnel on an ad hoc basis.

Finding: The inclusion of U.S. Coast Guard personnel, the Naval Criminal Investigative Service, or other law enforcement detachments or personnel on selected U.S. Navy ships could extend U.S. capabilities to respond to suspected smuggling or terrorist activities.

Recommendation 4: The Chief of Naval Operations should ask the Coast Guard, the Naval Criminal Investigative Service, or another law enforcement entity to provide legal personnel for selected U.S. Navy ships.

In order to realize theater engagement or Navy MSP goals, the USCG could be asked to forward-deploy additional vessels to specific areas of the world. These vessels would work for the COCOMs on missions accepted by the USCG. For instance, low-end USCG vessels might be the appropriate maritime component command for a military operation. The USCG's "sovereignty expertise" might be the right answer for the Navy/COCOM, allowing them to gain access that they could not otherwise obtain. Such actions could pave the way for greater trust and cooperation between countries, including between their military counterparts.

This activity and the associated program would affirm the concept of the USCG/Navy National Fleet. The USCG would need to be funded and staffed appropriately to take on this additional mission responsibility. A recent example was the use of a USCG cutter in the Gulf of Guinea. Likewise, USCG ships participated in PACOM activities on the Rim of the Pacific (RIMPAC) and deployments to Joint Task Force-150. Each year, COCOM requests for USCG vessels and training teams far outstrip the capacity of the USCG. In the Gulf of Guinea deployment, for instance, using low-end USCG vessels and the mission control center for a military operation was appropriate for the situation. It is the USCG sovereignty expertise, mentioned above, that many countries seek and that in turn give the Navy/COCOM access that would not otherwise be possible. Such actions can pave the way for greater trust and cooperation between one country and another and between their military counterparts. This would be true for all other USCG training teams that might be funded and made available to carry out specific missions.

Finding: The forward deployment of U.S. Coast Guard vessels can enhance and strengthen the engagement activities and thus increase the number of partnerships.

Recommendation 5: The Chief of Naval Operations should ask the Commandant of the Coast Guard to forward-deploy Coast Guard cutters to locations that offer opportunities for the joint enforcement of maritime security. These cutters would help to attain Navy and combatant commander engagement goals and would be the correct security assets to employ to meet theater cooperation goals.

Relatively speaking, the total effort required to expand the scope and depth of MSP is not large. Indeed, some of the overall funding can come from direct or in-kind contributions of the strategic partners themselves. MSP are based on the win-win concept—that is, they are of benefit both to relationships and to the flows of activity and information that sustain them. But at least for the initial period, the 1,000-ship Navy concept requires the Office of Management and Budget to scrutinize Navy programs and budgets not only to identify programs but also to include the funding needed for implementation of the MSP.

Finding: Secure, continuing funding is a key ingredient for sustaining and deepening maritime security partnerships.

Recommendation 6: To sustain and deepen maritime security partnerships (MSP) and to make such programs robust and stable, the Chief of Naval Operations should:

- Establish and assign to a specific office the coordination authority for programs and budgets for MSP in the Navy, throughout the Department of Defense (DOD), and across the federal agencies. This should include enhanced opportunities for professional education and for the necessary equipment and support services;
- Request that the Defense Security Cooperation Agency work with the State Department to significantly enhance the portfolio of international military education and training funds (e.g., those under Sections 1206 and 1208 of the National Defense Authorization Act of 2006, and COCOM Initiative Funds) for countries deemed key for MSP development. This activity—the implementation of a network of MSP—should also set a high priority on the institutionalization of an international legal training program;
- Task the Navy's International Programs Office to place high priority on funding the transfer of equipment, software, and services to support and intensify existing MSP and to develop new bilateral and multilateral MSP;
- Together with the appropriate officials at the State Department and other agency partners in MSP, request more funds for use by the maritime services, the State Department, and other relevant government agencies for training and

support of MSP initiatives or for activities at the International Maritime Organization and other relevant international organizations and multilateral frameworks to maintain and expand information-sharing programs and protocols;

- Propose to the appropriate parts of DOD the setting aside of a portion of research, development, test, and evaluation funds over the next 5 years to be committed under the Office of Manpower and Personnel guidance to the specific goal of improving technologies and techniques for easy, reliable information sharing and the continuous availability of common maritime operational pictures on as broad a basis as possible. These would subsume but go beyond the already programmed funding for MDA only that is now appropriated to the Office of Naval Research (see Chapter 3).

3

Information Sharing, a Key Enabler

MARITIME SECURITY

As discussed in the foregoing chapters, information collection and sharing are central to building trust; they also provide a basis for decisions and actions. In fact, the resulting transparency in and of itself arguably contributes to the maritime security of the United States and its partners. This chapter covers matters relating to the presence and activities of ships and craft on the surface of the oceans—from the high seas well into territorial waters. Such information helps us to understand—and therefore respond to—potential threats to maritime security. Also of interest is information on various cargoes, crew, the supply chain, and even ownership and management affiliations, which helps to identify illegal, suspicious, or threatening activities.

The Maritime Security Partnership Initiative

The committee believes that the formation of partnerships to improve maritime security is characterized by a number of fundamental principles:

- Maritime security around the globe will be advanced by strengthening existing partnerships and building new ones, with shared information the key enabler.
- It is envisioned that not only will action on the maritime security situation generally be accomplished at the regional or subregional level, but it will also have a collective global effect as well and will require some local improvements in the maritime security situation.
- It is in the interest of both the United States and its partner nations to share

information as widely as possible within regions and subregions and beyond, taking into account that threats to security often cross regional or subregional boundaries.

- The related objectives of extending reach and maximizing inclusiveness suggest that both the information to be shared and the system architecture for doing this involve unclassified information[1] and the use of commercial, Internet-based mechanisms.
- Beyond information collection and sharing—viewed here as having intrinsic value and serving as a fundamental building block when forging new partnerships—there is, of course, the matter of taking endgame action to deny or deter illegal or threatening activities.
- Improved information collection and sharing can be expected to generate a positive spiral in terms of increasingly effective coordinated action among maritime partners.
- The U.S. Navy, as one of the nation's main repositories of technical expertise and, often, the primary entity that interfaces with a potential partner entity, is well positioned to support the initiative on maritime security partnerships (MSP). The initiatives of the combatant commander (COCOM) and the Navy reflect the above fundamental principles.

After presenting some context and characterizing the current systems and capabilities, this chapter focuses on technical considerations—including architectures and technical options—for building and strengthening capability in three functional areas: sense/collect, analyze/fuse, and decide/act. Before new capabilities for maritime information collection and sharing can be shared with the partners, it will be necessary to agree on mutual responsibilities and obligations. Particularly in the case of nontraditional partners, it is the committee's view that some additional principles apply:

1. COCOM and Navy fleet experience has shown that the new partners are generally interested in local rather than regional or global maritime domain

---

[1]Pursuant to Executive Order 12958, classified information refers to official information that has been determined to require protection against unauthorized disclosure in the interest of national security and that has been so designated. Unclassified information refers to information that has not been determined to warrant classification; however, some unclassified information may be approved for public release whereas certain other information, such as International Traffic in Arms Regulations information, may not. Some maritime information that does not pertain to U.S. national security, such as Automatic Identification System reports, can be viewed as publicly available and therefore can be freely shared (subject only to constraints imposed by international agreements, such as IMO, as opposed to U.S. policy). When referring to such information, the U.S. Navy has coined the term "not classified," apparently to convey the notion of useful information sharing without the potential complexities of codified protection requirements. The term "unclassified," as used in this report, is viewed as encompassing "not classified" information.

awareness (MDA). Yet, this desire for local awareness can contribute to the larger global picture by enabling the early identification of vessels of interest.

2. When developing agreements on the sharing of information with members of the emerging partnership, it will be important to take into account the information that is currently available from local law enforcement (ports, supply chains, etc.) as well as information collected by technologically sophisticated surveillance systems. Each partner, including the United States, will have to consider what boundaries, if any, to place on the sharing of information, even unclassified information.

3. When seeking to strengthen and expand local surveillance capabilities, reaching agreement may well require a modest level of U.S. support and investment, ranging from technical support for the partner to obtaining permission to site our radar installation on the partner's sovereign territory.

Broadly speaking, the notion of a partnership usually entails reciprocity, which can take many different forms, including the exchange of money, information, and technical know-how.

Other Initiatives to Enhance Maritime Domain Awareness

The information that is being shared here is the kind that makes us and our partners aware of our maritime domain. MDA encompasses a growing spectrum of initiatives to develop capabilities, including the National Strategy for Maritime Security (NSMS) and the National Plan for Achieving MDA. As this committee was completing its efforts, the Department of Defense (DOD) was assigning to various federal agencies their responsibilities for MDA capability development and was securing the required funding. A memorandum from the Secretary of the Navy, dated May 17, 2007, called for an MDA "spiral 1" initial operational capability (IOC) by August 2008 for the U.S. Central Command (CENTCOM), the U.S. Pacific Command (PACOM), associated fleet elements, non-DOD U.S. organizations, and selected foreign partners in the western Pacific. On May 29, 2007, the Chief of Naval Operations (CNO) issued the document *Navy Maritime Domain Awareness Concept* to guide Navy efforts to improve MDA-related capabilities and develop related Fleet Concept of Operations (CONOPS).[2] A memorandum from the Deputy Secretary of Defense dated August 3, 2007, designated the Navy as DOD's executive agent for MDA and outlined the responsibilities and mechanisms for addressing requirements, investing resources, and supporting interagency efforts. This memorandum called for preparing a plan within 180 days to develop MDA capabilities. Additionally, DOD appointed a flag-level director of global maritime situation awareness (GMSA), a position that would

[2]Chief of Naval Operations (ADM Michael G. Mullen, USN). 2007. *Navy Maritime Domain Awareness Concept*, Department of the Navy, Washington, D.C., May 29.

complement the closely related, previously established position of director of global maritime intelligence integration (GMII).

Additionally, a national CONOPS for maritime domain awareness was published in August 2007, just as the committee was completing the draft of its report.[3] This CONOPS formally established the interagency GMSA office at the Coast Guard. The GMSA's current mission calls for it "to create a collaborative global, maritime, information sharing environment through unity of effort across entities with maritime interests." Although the committee did not have an opportunity to review this CONOPS, which was in response to the NSMS, the document apparently supports at least three notions that are elaborated on below from an MSP standpoint: (1) the importance of a modern, network-centric information technology (IT) capability for collecting, processing, and sharing information to support the MDA community; (2) the need for developing and managing an MDA information architecture to guide the evolution of this capability; and (3) the technical leadership that the Navy can and should exercise in this domain, presumably building on its role as executive agent for MDA within DOD, as noted above.

The committee notes, however, that the ongoing MDA-related initiatives identified above are largely focused on the analysis and dissemination of existing information and do not deal with the need for additional information from surveillance sensors, noted later in the section "Building Mission Capability."

The MDA efforts outlined above have of course been motivated by the U.S. commitment to implementing the maritime component of the global war on terror and addressing the associated homeland security and defense concerns.[4] This focus notwithstanding, it is clear that the issues being addressed (e.g., barriers to information sharing) and the capabilities being developed (e.g., improved vessel tracking) apply to the broader maritime security interests embodied in the MSP concept. The prosecution of MSP initiatives, then, is an outcome of the ongoing and emerging MDA efforts triggered by the earlier NSMS.

The findings and recommendations in this chapter are intended to advance

---

[3]See Emelie Rutherford, *Inside the Navy*, 2007, "CONOPS Finalized This Month: Metcalf Heads Up New Global Maritime Situational Awareness Office," August 20, pp. 1, 10; and *Inside the Navy*, 2007, "GMSA Office Will Target Policy Barriers: Challenges Cited in Sharing Data for New Maritime Awareness Effort," September 10, pp. 1, 9.

[4]The National Research Council's Naval Studies Board recently conducted a study on the role of naval forces in the global war on terror (GWOT; see NRC, 2007, *The Role of Naval Forces in the Global War on Terror: Abbreviated Version*, National Academy Press, Washington, D.C.). Background information pertaining to the origins of the term "GWOT" can be found in documents such as (1) The White House (George W. Bush), 2006, *The National Security Strategy of the United States of America,* Washington, D.C., March, p. 12; (2) Office of the Chairman, Joints Chiefs of Staff, 2006, *National Military Strategic Plan for the War on Terrorism,* Washington, D.C., February 1, p. 3; and (3) Secretary of Defense, 2006, *Quadrennial Defense Review Report,* Department of Defense, Washington, D.C., February 6. The NRC Committee on the "1,000-ship Navy"—A Distributed and Global Maritime Network saw its charter as being neither to endorse nor to replace the term "GWOT."

the MSP concept by leveraging, complementing, and in some cases extending the broader U.S. MDA efforts.

OPERATIONAL MODELS

As elaborated in the Chapter 2 review of existing and emerging international partnerships, "one size does not fit all" when it comes to information-sharing arrangements and the enabling technical mechanisms. Differences are traceable to a number of factors:

- Different levels of trust,
- The distinction between bilateral and multilateral arrangements,
- A focus on coordinated action at the tactical level rather than on information sharing, and
- Uneven levels of technological maturity and sophistication.

Much of the current activity is associated with the burgeoning of automatic identification systems (AISs) on all commercial ships over 300 gross tons (GT) and on U.S. Navy ships. Figure 3.1 is a modified version of Figure 2.1 (which

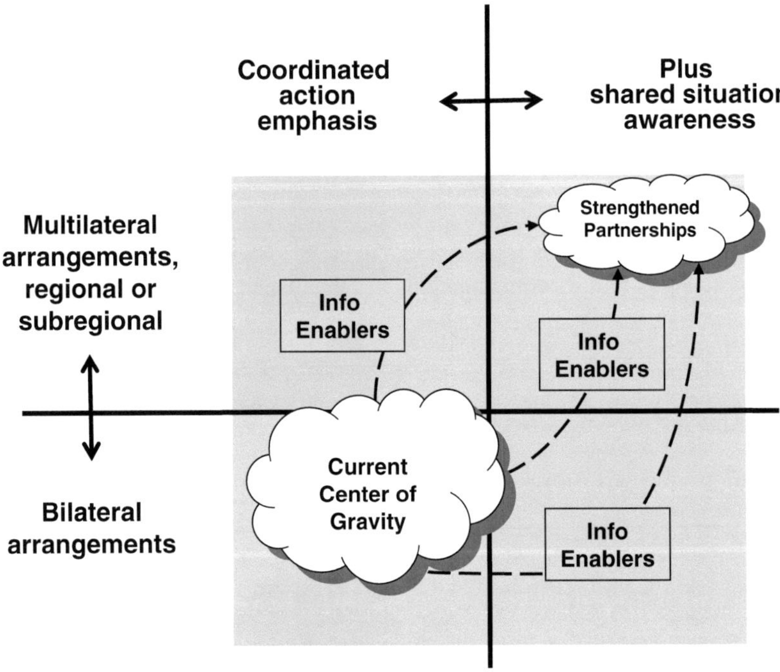

FIGURE 3.1 Current and emerging international maritime security partnerships.

characterizes the nature of existing or emerging partnership arrangements). The modifications are intended to highlight the role of "information enablers" (both the information content and the systems capabilities) as a foundation for effective partnerships. These enablers would support both information sharing to gain situation awareness and subsequent coordinated action. As depicted, the center of gravity of current maritime partnerships resides in bilateral arrangements focused on the coordinated execution of tactical actions such as interdiction that support common security interests.

Figure 3.2 depicts an example of the sharing of information referred to in the upper-right quadrant of Figure 3.1. It shows the position and movement of vessels around the island nation of Singapore, reflecting the merging of information broadcast automatically by ships that comply with international AIS standards and data obtained from coastal radar installations.

Figure 3.2 suggests how the sharing and combining of particular sets of information could enable coordinated multilateral or bilateral action. Later sections of this chapter explore these enablers, and the committee then develops some findings and recommendations regarding their conceptualization, design, and implementation. It is noted here, and elaborated on below, that activities being carried out by the Navy and the larger maritime security community represent substantial initiatives to advance these enablers. The committee's aim is to refine the original 1,000-ship concept and thereby contribute to further progress.

CURRENT AND EMERGING INFORMATION ARCHITECTURES

Not surprisingly, having a range of information architectures allows the sharing of information among maritime partners, from mature partnerships among alliance members—for example, the North Atlantic Treaty Organization [NATO])—through temporary coalitions formed for a specific mission purpose (e.g., Joint Task Force-150 supporting operations in Iraq), to less mature and often more ad hoc arrangements with "nontraditional" partners (e.g., the Gulf of Guinea Initiative). It is instructive to review existing and emerging information-sharing systems and networks and to identify their fundamental architectural characteristics.

Some regions have already established networks to share MDA information. For example, the Malacca Strait Initiative partnering Singapore, Indonesia, and Malaysia is already operational; the Gulf of Guinea network, still in its formative stage, has generated a great deal of interest on the part of the potential partners; and the Joint Interagency Task Force-South (JIATF-S), addressing drugs and other law enforcement concerns in the Caribbean region, is functioning effectively. However, while many capabilities support MDA systems around the world, they are a patchwork of efforts. There is no overarching MDA architecture. With the exception of the International Maritime Organization (IMO)-sanctioned AIS and the Long-Range Identification and Tracking (LRIT) reporting systems for

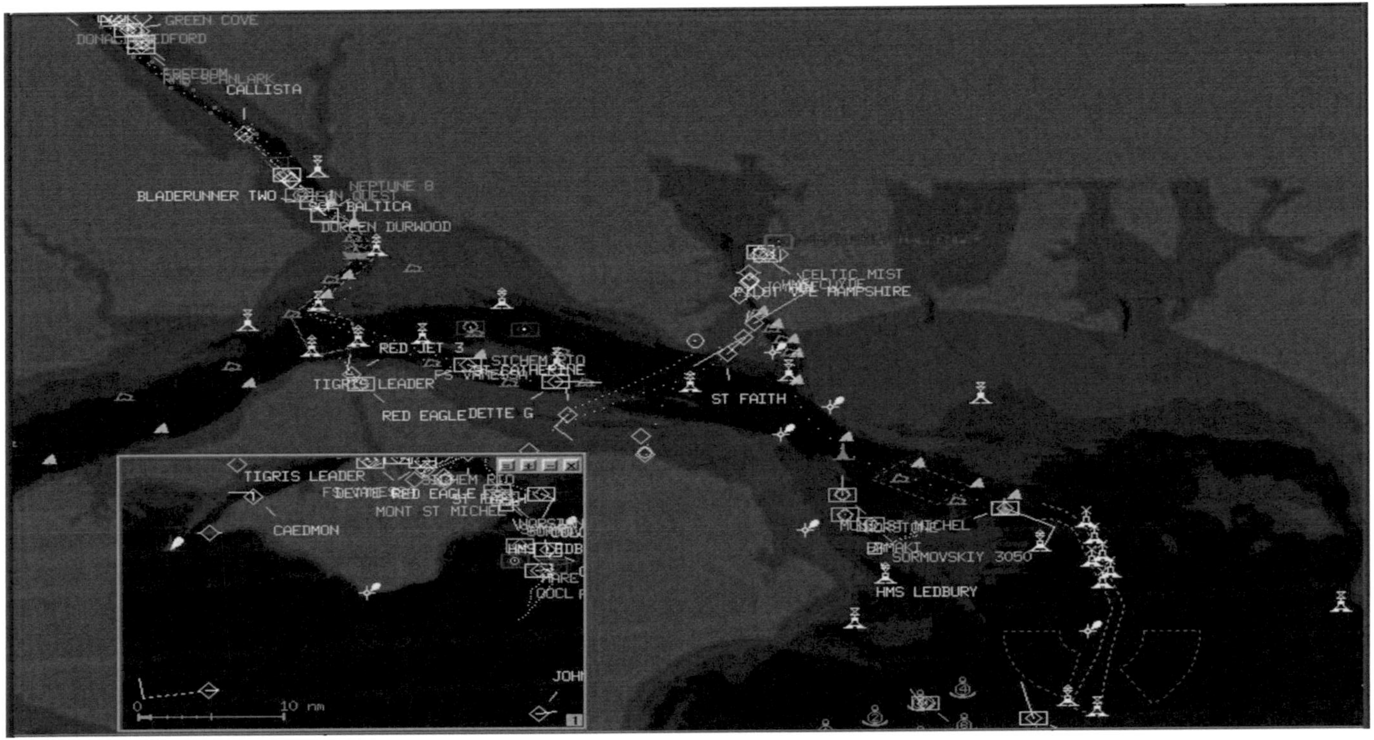

FIGURE 3.2 Singapore area maritime "picture." SOURCE: COL James Soon, Republic of Singapore Navy, Head, Defence Technology Office, Embassy of Singapore, "The 1,000 Ship Navy: A Perspective from Singapore," presentation to the committee, Washington, D.C., March 14, 2007.

commercial ships, current arrangements for sharing MDA information, though sometimes multilateral, are mostly inefficient and lack broad application.

It will take considerable effort to coordinate all the existing capabilities, extend them, and disseminate information on a timely basis to those maritime law enforcement organizations that can take the appropriate action while still respecting commercial and national sensitivities and proprietary interests. Mobilizing the U.S. government to assist other nations in creating more comprehensive MDA and enlisting, connecting, and sustaining the capabilities of the maritime law enforcement organizations will be a long, continuing process. At the same time, this process would build trust and transparency with other nations, contributing substantially to global cooperation.

The unifying concept behind maritime security partnerships is information sharing. Using the vocabulary that has been adopted by the U.S. initiatives responding to the NSMS, the information to be shared is referred to as MDA. Because a more comprehensive MDA system would facilitate the identification of threatening activities and anomalous behavior, it would be useful for the U.S. government, encouraged by the CNO, to devote additional effort to the collection, analysis, and distribution of maritime domain awareness information and to support the development of regional partnerships that could mount a concerted response to regional threats.

Current Systems for Sharing Information

Table 3.1 summarizes seven representative systems selected because (1) they specialize in the sharing of maritime domain information and (2) they span a spectrum of kinds of information challenges, from Secret to unclassified. The table covers a variety of systems, from operational networks that facilitate the sharing of Secret information among both traditional alliance and coalition maritime partners in Iraq (Joint Task Force-150 CENTRIXS) to emerging demonstration networks for the sharing of unclassified, commercially available AIS (and other) information with nontraditional partners (such as the U.S. Naval Forces, Europe (NAVEUR)-led Gulf of Guinea Initiative). Noting the positive characteristics of the Regional Maritime Awareness Capability (RMAC) and Comprehensive Maritime Awareness (CMA) Joint Concept Technology Demonstrations (JCTDs) as well as some differences in approach between them, the committee strongly endorses the notion of regional pilots—generally led by the COCOMs and supported by the associated fleet elements as a pragmatic way to make progress while building fundamental relationships. It would seem that a maritime pilot involving the northeast African coastal nations might warrant consideration as the new AFRICOM begins to undertake outreach. Further descriptive information for each of the seven systems follows.

TABLE 3.1 Current and Emerging Information-Sharing Systems

Systems and Initiative	Lead Organizations	Status	Users	Information Shared	Communications
CENTRIXS	DISA MNIS JPO	Fielded	5 COCOMs, 77 nations, and NATO; all U.S. Navy ships	Releasable Secret COP, e-mail, chat	Dedicated nets
CNIES	U.S. Southern Command	Fielded	JIATF-S	Unclassified COP, e-mail, chat	Internet
MSSIS	U.S. Navy Sixth Fleet	Fielded	U.S. Navy, other navies, NATO (26 countries)	Automatic Identification System (AIS) (identification, position, other)	Internet
NAIS	U.S. Coast Guard	Increment 1 (IOC) October 2007	U.S. Coast Guard	AIS	Department of Homeland Security net

RMAC JCTD	U.S. European Command	Demonstration	Demonstration	AIS, other sensor data (radar)	VHF/UHF radio, cell phones, Internet
CMA JCTD	COMPAC (PACFLT), USNORTHCOM, COMNAVEUR, C6F, U.S. Coast Guard	Demonstration	Demonstration	SCI, GENSER, unclassified, coalition releasable	JWICS/SIPRNET/NIPRNET/Internet
LRIT	International Maritime Organization	IOC December 2008	Flag states, port states, coastal states	Ship identification, position, date/time	Commercial COMSAT, Internet

NOTE: CENTRIXS, Combined Enterprise Regional Information Exchange System; CNIES, Cooperating Nations Information Exchange System; MSSIS, Maritime Safety and Security Information System; NAIS, Nationwide Automatic Identification System; RMAC JCTD, Regional Maritime Awareness Capability Joint Concept Technology Demonstration; CMA JCTD, Comprehensive Maritime Awareness Joint Concept Technology Demonstration; LRIT, Long-Range Identification and Tracking; DISA MNIS JPO, Defense Information Systems Agency Multinational Information Sharing Joint Program Office; COMPAC, Commander, Pacific; PACFLT, U.S. Pacific Fleet; USNORTHCOM, U.S. Northern Command; COMNAVEUR; Commander, Naval Forces Europe; C6F, Commander Sixth Fleet; IOC, initial operational capability; COCOM, combatant commander; NATO, North Atlantic Treaty Organization; JIATF-S, Joint Interagency Task Force-South; COP, common operational picture; SCI, sensitive compartmented information; GENSER, General Service; VHF, very high frequency; UHF, ultrahigh frequency; JWICS, Joint Worldwide Intelligence Communications System; SIPRNET, Secret (formerly Secure) Internet Protocol Router Network; NIPRNET, Nonclassified Internet Protocol Router Network; COMSAT, communications satellite.

Combined Enterprise Regional Information Exchange System

CENTRIXS is a combination of separate multilateral and bilateral government networks. Key CENTRIXS networks include the Global Terrorism Task Force (GTTF) network (supporting Operation Enduring Freedom, 66 nations) and the Multinational Coalition Forces–Iraq (MCF–I) network (51 nations). Five combatant commands (COCOMs) are CENTRIXS-enabled, and there are 77 participating nations plus NATO, 11 bilateral agreements, and over 26,000 users. CENTRIXS evolved from various networking initiatives developed by the COCOMs to meet their regional information exchange needs. Although there are many individual CENTRIXS networks, they are now centrally supported and managed by the Joint Program Office's (JPO's) Multinational Information Sharing (MNIS) under the Defense Information Systems Agency (DISA).

CENTRIXS is Web-centric and employs both commercial off-the-shelf (COTS) and releasable government off-the-shelf (GOTS) products. It includes MS Office automation tools, the GOTS command and control personal computer (C2PC) tool for situation awareness display, collaboration tools, and the GOTS integrated imagery and intelligence (I3) tool. A CENTRIXS workstation user is able to access browser-based products and databases, receive and display non-real-time track data feeds on a map background, send e-mail with attachments, and conduct collaboration sessions.[5]

While CENTRIXS provides significant operational capability and has become an essential tool for conducting current operations, areas for improvement have been identified and are being worked on. According to CENTCOM, ". . . inconsistencies in data owner guidance from various producers, a lack of manageable technical solutions, and a cumbersome accreditation and certification process have combined to frustrate seamless data dissemination via electronic (such as CENTRIXS) networks. These problems have directly contributed to the proliferation of multiple separate networks. The burden of additional networks has consumed limited resources and manpower and imposed an opportunity cost on CENTCOM's coalition warfighting efforts."[6] The MNIS JPO has initiatives under way to address many of these issues, but this is clearly an area that needs continuing focus.

The DISA MNIS JPO is implementing a plan to centralize CENTRIXS service provision at the Defense Enterprise Computing Centers (DECCs) in Columbus, Ohio, and Hawaii. The MNIS JPO also manages and supports the

---

[5]Jill L. Boardman, Lockheed Martin Information Technologies, and Donald W. Shuey, Department of the Air Force, U.S. Central Command (CENTCOM). 2004. "Combined Enterprise Regional Information Exchange System (CENTRIXS); Supporting Coalition Warfare World-Wide," CENTCOM, MacDill Air Force Base, Fla., April, p. 13.

[6]Jill L. Boardman, Lockheed Martin Information Technologies, and Donald W. Shuey, Department of the Air Force, U.S. Central Command (CENTCOM). 2004. "Combined Enterprise Regional Information Exchange System (CENTRIXS); Supporting Coalition Warfare World-Wide," CENTCOM, MacDill Air Force Base, Fla., April, p. 12.

Globally Reaching Interactive Fully Functional Information Network (GRIFFIN), supporting classified information sharing and collaboration with and among the United Kingdom, Australia, Canada, and New Zealand.

Cooperating Nations Information Exchange System

The Cooperating Nations Information Exchange System (CNIES) is used by JIATF-S and 11 cooperating nations in South and Central America to suppress illicit maritime drug traffic. The 11 include European nations with naval operations in the Caribbean basin. JIATF-S is staffed with personnel from the Departments of Defense, Homeland Security (USCG), Justice (Drug Enforcement Administration, Federal Bureau of Investigation), and Treasury (U.S. Customs and Border Protection). JIATF-S is currently commanded by a USCG flag officer and reports to SOUTHCOM. Its mission is to counter illicit trafficking operations, to promote security cooperation, and to coordinate country team and partner nation initiatives in order to defeat the illicit flow. This mission was expanded after 9/11 to explore the linkage between drug trafficking and terrorism.

Cooperating nations gain access to CNIES by entering into bilateral agreements with the United States. These agreements are negotiated through the U.S. Department of State in the context of United Nations conventions, including the 1988 United Nations Convention Against Illicit Traffic in Narcotic Drugs and Psychotropic Substances. The agreements describe procedures for the conduct of counterdrug operations in the waters, territories, and airspaces of the participating nations. These procedures include provisions similar to those summarized in Table 3.1 for ship riders, personnel from one nation who embark on a vessel belonging to another nation and who can authorize the boarded vessel to assist in the enforcment of the laws of their nation.

CNIES extracts and distributes portions of SOUTHCOM classified common operational pictures (COPs) for South and Central America (both air and surface tracks) to each cooperating nation according to the respective bilateral agreements. Geographic filtering is used to give each nation a different picture, with 11 versions of the COP in all. The picture is displayed using "releasable GOTS" client software provided by the U.S. C2PC tool. Cooperating nations can add tracks to their operational picture, but these tracks do not affect the U.S. COPs and are visible only within the CNIES domain. An extensive network of U.S. over-the-horizon (OTH) radars and some cooperating nation radars provides persistent air surveillance of the drug transit zone, but surface radar surveillance is much more limited. The primary use of the COP is to coordinate drug interdiction operations. For this reason, there is little emphasis in this theater on the acquisition or use of AIS data, because little drug trafficking is associated with registered commercial shipping.

The CNIES is based on commercial Internet technology. A Radiant Mercury guard is used to strip classified data from the U.S. version of the COP. The portion of the COP provided to a particular nation is maintained on hardware dedicated to that nation. Commercial products are used to establish a virtual private network with each cooperating nation, and commercial firewalls and routers are used for information security. In addition to enabling the sharing of COPs, CNIES ensures e-mail and chat with automatic translation.

In addition to technical capabilities, CNIES includes liaison officers assigned to JIATF-S from the other cooperating nations. These liaison officers provide the face-to-face contact that is essential in planning and conducting drug interdiction operations. Particularly sensitive information is generally handled by voice communication between individuals with trusted relationships.

Maritime Safety and Security Information System

The Maritime Safety and Security Information System (MSSIS), currently in use in the Mediterranean, Europe, and Africa, was conceived by the Department of Transportation's (DOT's) Volpe Center and the U.S. Navy's Sixth Fleet as an unclassified, freely shareable MDA network. MSSIS is also a multinational, Internet-based network primarily for sharing real-time AIS data derived from shoreside, waterborne, and airborne platforms. The received information is centrally processed on a server at the U.S. DOT Volpe Center using software developed by Volpe for this purpose. The resulting plots and associated information tags are then made available via protected Internet access (Secure Sockets Layer [SSL]/password protection) to the contributing sites and the headquarters, planning, and response organizations.

MSSIS implementation in the field requires Internet connection, an AIS receiver, and a laptop PC running the Volpe Center's TransView (TV32) geographic information system software. TV32 is configurable to satisfy a range of display requirements, including enhanced navigation safety, waterway efficiency, traffic situation awareness, force protection, and data analysis. A client user can interrogate the MSSIS Web site by geography but needs to run the Volpe TV32 client software to get at the AIS data for the region.

One important goal of MSSIS is to support the RMAC JCTD Gulf of Guinea Initiative, described below, to help and encourage littoral nations to better monitor and police their seaward approaches, with a goal of reducing poaching and piracy. An effort is just starting to establish metrics applicable to this effort—for instance, the fraction of time that an area is under effective surveillance. The current supplemental budget request to Congress includes Section 1206 funds for the U.S. Navy to buy AIS equipment for some Gulf of Guinea states. One problem is that many coastal state government facilities have little or no Internet connectivity. Further, only limited attention has been paid to date to the need to overlay AIS data with radar surveillance data, where available.

MSSIS is currently feeding data to various U.S. and partner organizations, including research agencies such as the Defense Advanced Research Projects

Agency (DARPA), operational elements in the Mediterranean, and related demonstration initiatives (such as CMA JCTD).

Nationwide Automatic Identification System

The Nationwide Automatic Identification System (NAIS) is being developed by the USCG to enhance maritime safety, security, and mobility. NAIS will augment current capabilties to receive, distribute, and utilize AIS data. NAIS is being developed in three increments:

- Increment 1, AIS receive in critical ports and coastal areas,
- Increment 2, AIS receive and transmit nationwide, and
- Increment 3, long-range (2,000 nmi) AIS receive.

NAIS IOC is scheduled for October 2007, and final operational capability (FOC) is scheduled for October 2013. AIS is intended to improve the safety of navigation by providing:

- A ship-to-ship mode for collision avoidance,
- A means for littoral states to obtain information about a ship and its cargo, and
- A vessel traffic services (VTS) tool.

As will be the case for the new LRIT system, the International Convention for the Safety of Life at Sea (SOLAS) already requires AIS for certain classes of ships (including ships of 300 GT or more[7]), and IMO has developed performance standards for AIS, primarily to help prevent collisions. These standards require that ships broadcast their identity, position, speed, and heading and other information. The reporting interval depends on ship speed and maneuvering and can be as short as 2 sec. Each shipboard AIS system consists of one very high frequency (VHF) transmitter, two VHF time division multiple access (TDMA) receivers, one VHF digital selective calling (DSC) receiver, and standard marine electronic communications links to shipboard display and sensor systems. AIS uses self-organizing TDMA to handle over 4,500 reports per minute. Range depends on the transmitter and receiver antenna height; a typical value at sea is 20 nmi.

The use of AIS information for other than local collision avoidance purposes depends entirely on the existence of, and distance to, equipment that can receive the AIS transmissions from the ships and direct the information to processing centers that can combine it with information from other sources and assess its implications for security. Many coastal nations are already well along in the installation

[7]The USCG is considering requiring the AIS carriage on vessels smaller than 300 GT, including pleasure craft, tugs, barges, and so on.

of coastal AIS receivers and the integration of the received information with that available from coastal radars. The absence of conforming AIS information from a large ship being tracked by radar could be considered suspicious. On the other hand, analysis of AIS information in areas without radar coverage is much more problematic. The NAIS program is concentrating initially on the approaches to ports that already have considerable radar surveillance in place or in train.

NAIS Increment 1 is increasing AIS coverage to 55 critical U.S. ports and 9 U.S. coastal areas. In addition, a storage, correlation, and dissemination capability at the USCG Operations Systems Center and a management and monitoring capability at the USCG are being established. AIS data are being fed to the USCG COP and the Maritime Awareness Global Network (MAGNet) and are available to other users via an AIS Web service.

NAIS Increment 2 will provide nationwide coastal AIS receive coverage out to 50 nmi and transmit coverage out to 24 nmi. It will implement a service-oriented, network-centric architecture that provides data dissemination services to all maritime stakeholders.

NAIS Increment 3 will extend AIS receive coverage out to 2,000 nmi. To achieve this capability, the USCG is investigating approaches such as these:

- AIS-equipped low-Earth-orbiting satellites,
- AIS-equipped offshore platforms and buoys using commercial satellite communications, and
- AIS-equipped aircraft and ships (USCG, Navy, and commercial).

If the extension of U.S. AIS coverage well beyond the available radar coverage is to be operationally useful, it will require the development and employment of sophisticated anomaly detection techniques, as described in a later section of this chapter.

Regional Maritime Awareness Capability Joint Capability Technology Demonstration

The Regional Maritime Awareness Capability (RMAC) Joint Capability Technology Demonstration (JCTD) is an international program sponsored by the U.S. European Command (EUCOM) and the U.S. Office of the Secretary of Defense (OSD). The RMAC JCTD provides an MDA capability for the understanding of maritime activities that impact regional and international safety, security, economics, and environment primarily in the Gulf of Guinea. By integrating off-the-shelf maritime sensors, communications systems, and software, the RMAC system will allow detecting, tracking, identifying, displaying, and sharing information about surface vessels at least 20 meters long between 10 and 25 nmi from ports, harbors, and critical assets.

The RMAC JCTD employs COTS sensors (radar, electro-optical infrared,

AIS, binoculars), COTS computers (Windows boxes, Solaris boxes), unclassified GOTS software (components of SureTrak, tactically integrated sensors [TISs], and TV32), commercial security technology, and TIS service-oriented architecture for publishing and subscribing information. RMAC can communicate using Link 11 or Link 16. Designed to provide maritime surveillance capability across a spectrum of coalition partners (from simple to sophisticated versions), RMAC is now set up in São Tomé and was planned for installation in Nigeria by the end of the summer in 2007. This capability will be operated, maintained, and sustained in a manner that fosters local ownership of regionally and internationally shared maritime security assets. The Department of State is an important player in this effort.

Comprehensive Maritime Awareness Joint Concept Technology Demonstration

The objective of the Comprehensive Maritime Awareness (CMA) Joint Concept Technology Demonstration (JCTD)[8] is to improve maritime security by acquiring, integrating, and exchanging relevant maritime activity information on regional threats and focusing limited interdiction and inspection assets on the most probable threats. Participants include PACOM, NORTHCOM, and EUCOM. Singapore is an international partner. The Naval Research Laboratory is the technical manager.

The technical focus of the CMA JCTD includes the development and demonstration of the importance of information sharing for improved maritime awareness—both interagency sharing and international sharing—along with demonstrating improved information management techniques, such as application of the DOD net-centric data strategy (see Figure 3.3).

In addition, to cope with the large volume of maritime information to be made available under MDA intitiatives, the CMA JCTD is developing and integrating automatic tools to provide timely and accurate maritime situational awareness, to identify and prioritize relevant and actionable information, and to acquire, fuse, and manage disparate information.[9] The CMA JCTD emphasizes the exchange of classified information, offering the requisite operational benefits but also introducing information protection requirements that are not fully compatible with a keep-it-simple, low-cost-of-entry approach to information sharing with nontraditional partners.

The CMA JCTD is being conducted in three spirals, with a demonstration at the end of each spiral:

---

[8]A JCTD is a DOD program to rapidly move advanced technology into the hands of warfighters in the field.

[9]Chris Dwyer, Naval Research Laboratory. 2007. "Comprehensive Maritime Awareness (CMA) Joint Capabilities Technology Demonstration (JCTD)," *Proceedings of SPIE* [Society of Photo-Optical Instrumentation Engineers], Vol. 6578, Defense Transformation and Net-Centric Systems 2007 [Conference], Orlando, Fla., April 9-13.

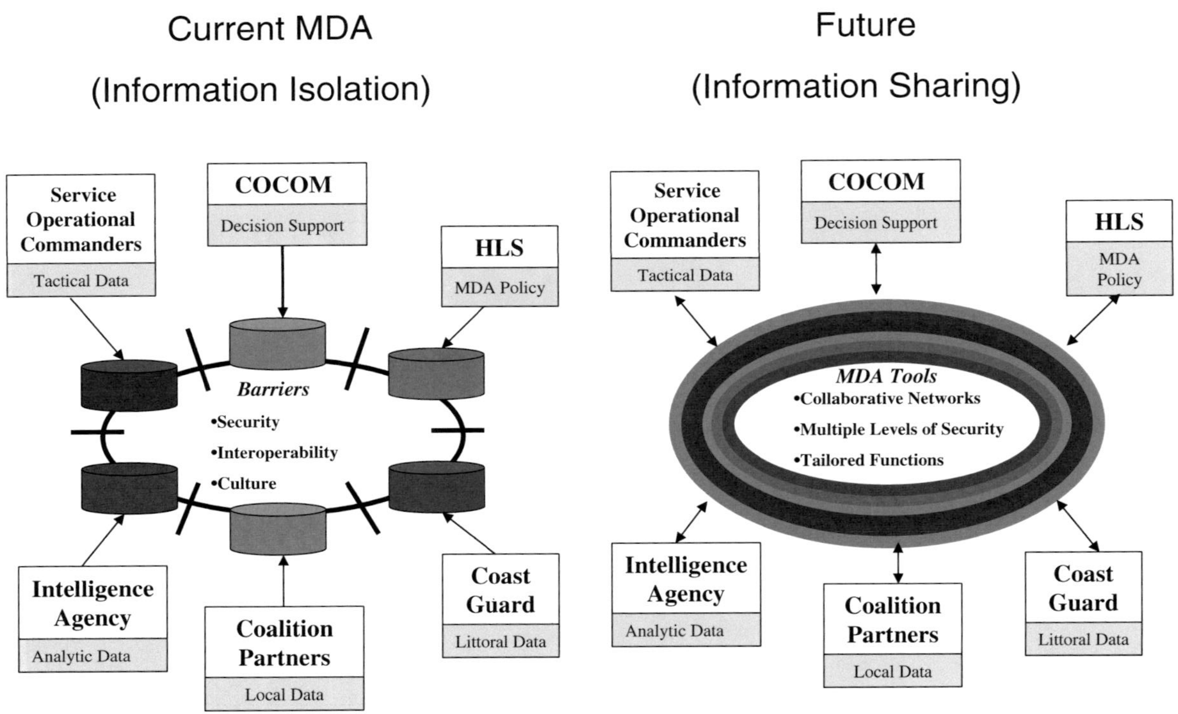

FIGURE 3.3 The CMA JCTD: Demonstrating improved information sharing and management. NOTE: HLS, homeland security. SOURCE: Chris Dwyer, Naval Research Laboratory. 2007. "Comprehensive Maritime Awareness (CMA) Joint Capabilities Technology Demonstration (JCTD)," *Proceedings of SPIE* [Society of Photo-Optical Instrumentation Engineers], Vol. 6578, Defense Transformation and Net-Centric Systems 2007 [conference], Orlando, Fla., April 9-13.

• *Demonstration 1.* Communications pipe between the regional operating centers and selected COCOMs (December 2006).

• *Demonstration 2.* Common distributed virtual database/information extraction (CDVD/IE) and other integrated capabilities and technologies across the participating COCOMs; selected U.S. federal, state, and local government entities; and coalition partners (fall 2007).

• *Demonstration 3.* Demonstration of a net-centric interagency exchange network based on service-oriented architecture technologies (fall 2008).

Long-Range Identification and Tracking System

The LRIT system (Figure 3.4) is being developed under the auspices of the IMO, an agency of the United Nations concerned with safety, environmental concerns, legal matters, technical cooperation, maritime security, and the efficiency of shipping. LRIT is being implemented under the authority of the International Convention on the Safety of Lives at Sea (SOLAS) for security and search-and-rescue (SAR) purposes. The SOLAS regulation on LRIT does not create or affirm any new rights of states over ships beyond those already existing in international law. The transmission of LRIT information is intended to be operational by December 31, 2008.

Ships subject to SOLAS (including cargo ships of 300 GT and up on international voyages as well as several other categories of vessels) will be required to transmit their identity, position, and the date and time of the position hourly. This information can readily be transmitted using current shipboard Global Maritime Distress and Safety System equipment at a cost of about 50 cents per transmission. Each ship will transmit its information to a data center specified by its flag state using services provided by communications service providers—for example, the International Maritime Satellite and Applications Service Providers. The data centers may be national, regional, cooperative, or international and may be associated with a Vessel Monitoring System. Using a data distribution plan and international routing rules established under the auspices of the IMO, this information will be provided to flag states, port states,[10] and coastal states[11] and for use in SAR. The Internet will be used where available. It should be noted that the resulting information, though consolidated and disseminated as shown, is not uniformly and freely shared among all using parties.

It should also be noted that LRIT is being developed in parallel with U.S. efforts to demonstrate the utility of communications satellites to relay AIS from existing ships' equipment when they are beyond the range of shoreside receiv-

---

[10]A port state has the right to LRIT information for a ship that intends to enter a port facility, at a distance or time set by the port state, but not in internal waters of another contracting government.

[11]A coastal state has the right to LRIT information for all ships, regardless of flag, within 1,000 nmi of the coast, but not in internal waters of another contracting government or in the territorial sea of the contracting government whose flag the ship is entitled to fly.

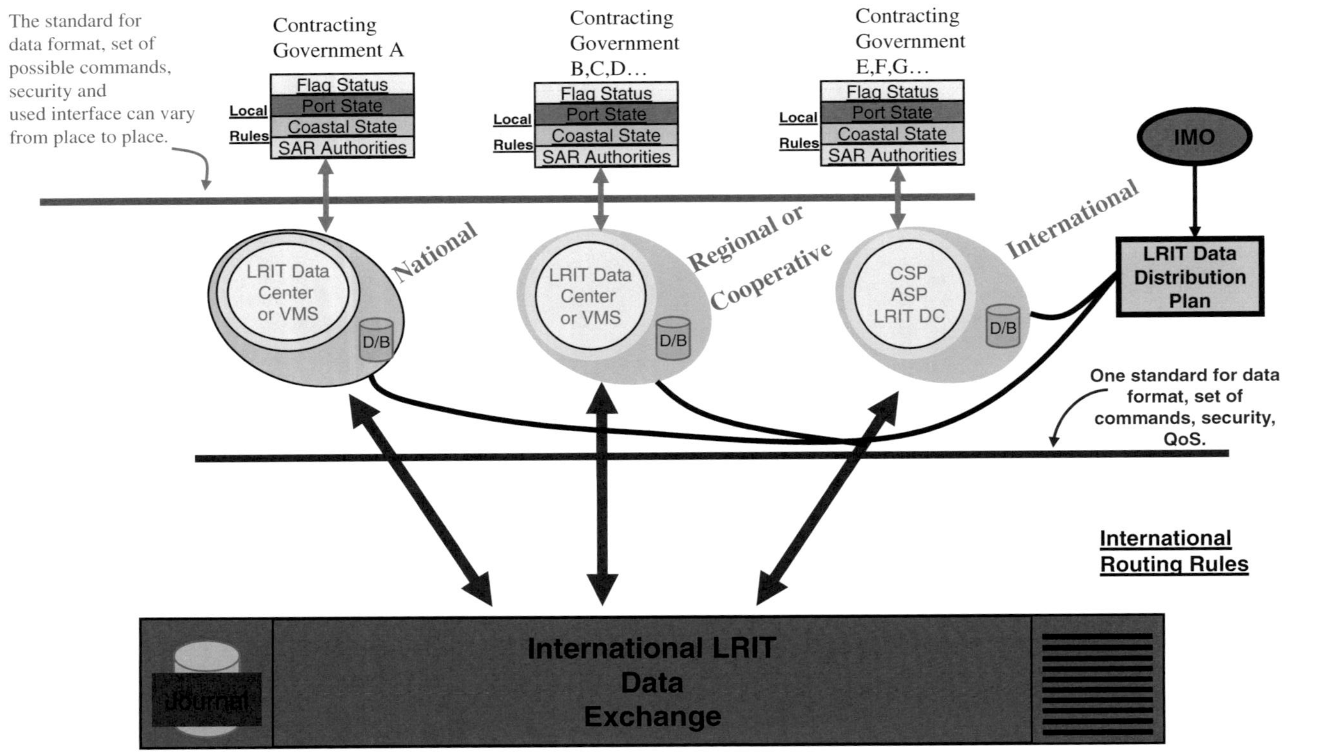

FIGURE 3.4 LRIT system. SOURCE: Chris Trelawny, Head, Maritime Security Section, International Maritime Organization, "IMO Perspective," presentation to the committee, Washington, D.C., February 7, 2007. For definitions of acronyms, see Appendix G.

ers. If such AIS global connectivity can be achieved, the need for LRIT would logically decline.

Other Related Navy Maritime Domain Awareness Initiatives

The systems described in Table 3.1 are good examples of existing technology and systems that are being leveraged to enable maritime information sharing. Beyond these systems and demonstration initiatives, the Navy—specifically, the Deputy Chief of Naval Operations for Communication Networks (N6) and the Program Executive Office (PEO) for Command, Control, Communications, Computers, and Intelligence (C4I)—has undertaken to develop a maritime information-sharing architecture that can be applied to sharing public and/or unclassified information with nontraditional partners.

The Navy, working with the USCG (CG-6), has taken the lead in requirements analysis for maritime domain collaboration and information sharing. Functional capabilities have been identified in the areas of (1) connectivity and reach-back, (2) interoperability tools, (3) collaboration tools, (4) data aggregation, (5) display/visualization, (6) correlation/fusion, and (7) cross-domain information sharing, including multilevel security and multinational information sharing. Given these requirements, the N6 and the PEO for C4I have launched (as of this writing) an investigation into a large set of MDA-related technologies and initiatives with the intent to develop a prototype system that leverages commercial software tools and select DOD research to fill technology gaps. The Navy (N6 lead) undertook a short-turnaround (18-month) effort, which will result in a prototype solution, demonstrating an architecture that should be considered as a starting point for implementing many initiatives. More recently, as noted above, the Secretary of the Navy has directed development of a spiral 1 MDA capability that builds on the N6 efforts reviewed by this committee.

N6 efforts to leverage the MDA data sharing (DS) community of interest (COI) activity and the related prototyping effort deserve particular mention. The MDA DS COI aims to transform data discovery and access from stovepiped systems to Web services using commercial networking technology and the MDA COI data standards. The fundamental issue of data interoperability is being addressed in this MDA initiative as part of a broader Office of the Secretary of Defense/Networks and Information Integration assault on data interoperability, an issue that will become central as the nature and scope of shared information content broaden in an MSP context.

Ongoing efforts (as of this writing) focus on the use of commercial technology in general and on the use of commercial information protection technology in particular to share information. Three information bins implying different levels of protection are depicted in Figure 3.5, with the Navy focused on unclassified information.

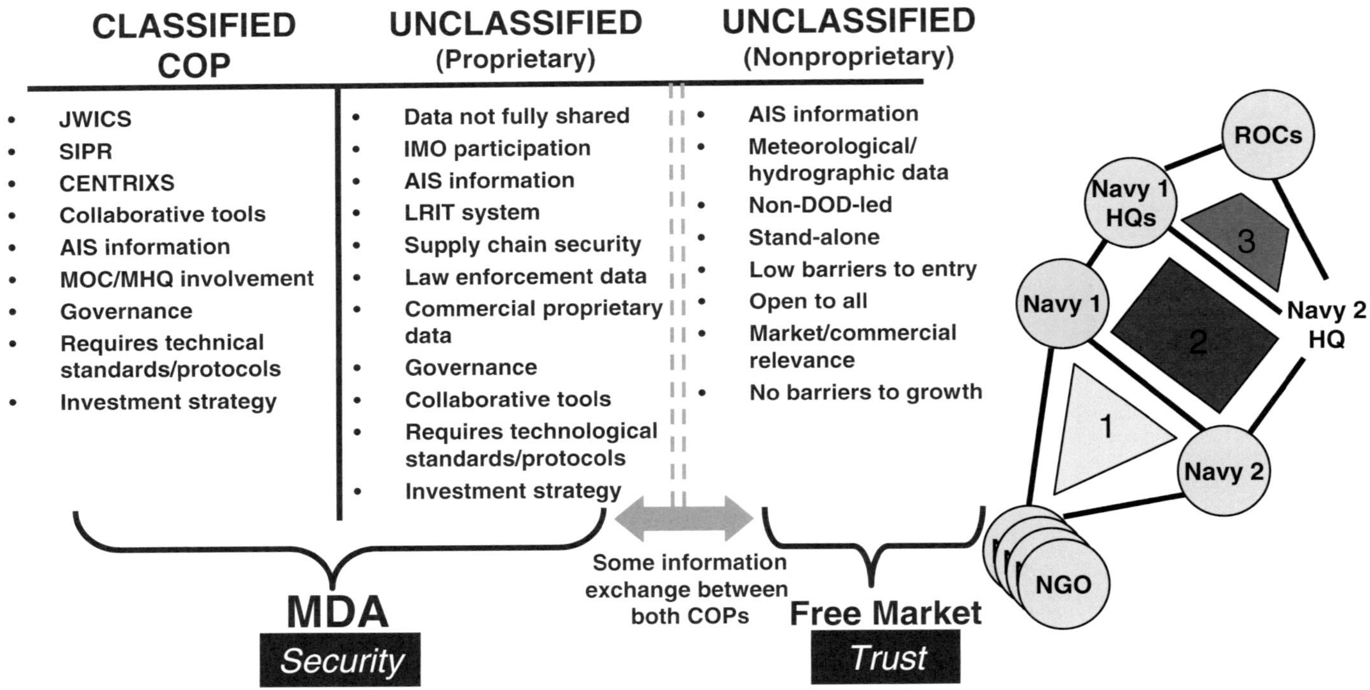

FIGURE 3.5 MDA bins for data sharing. NOTE: JWICS, Joint Worldwide Intelligence Communications System; SIPR, Secure Internet Protocol Router; MOC/MHQ, Maritime Operations Center/Maritime Headquarters; ROC, Regional Operations Center. See also Appendix G. SOURCE: RADM Kenneth Deutsch, USN, Director, Warfare Integration, Office of the Chief of Naval Operations (N6F/N83), "National Academies Naval Studies Board," presentation to the committee, Washington, D.C., January 10, 2007.

The thrust of the Navy effort and that of the related MDA DS COI effort—along with the management steps outlined above—are generally applauded. Note, however, that this committee is recommending serious consideration of additional capabilities, particularly in the area of promising surveillance options, as elaborated in subsequent sections.

Observations

Architectural Commonalities

Despite differences across the range of architectures described above, one can identify common themes and elements that appropriately reflect the pervasive adoption of modern, commercially based, Internet-like architectures. These are especially important in order to provide inherently low-cost, low-risk interoperability and commonality. Common themes reflect network-centric architectural principles and attributes, including these:

- Networking based on applying the Internet Protocol as an interoperable mechanism for exchanging data;
- Use of commercially based Web technologies and products for exploiting the IP-based networking:
  —Web-browser-based access to applications and data, and
  —Development of common MDA COI vocabularies that can be represented in flexible markup languages like extensible markup language; and
- Use of simple viewers for data presentation (sometimes releasable GOTS rather than COTS, with CNIES (JIATF-S) and the MSSIS TV32 viewers as cases in point).

Tailoring for the Nontraditional Partner Case

The higher-end CENTRIXS architectures provide substantial capability and, as elaborated above, are supporting critical coalition operations today. However, these architectures have features that violate the low-cost-of-entry and keep-it-simple principles when starting from scratch to build trust with nontraditional partners:

- Risk and complexity associated with the sharing of classified information;
- Reliance on government-developed software (versus COTS), which may also add technology-sharing issues and demand more operator training for the partner's personnel; and
- Reliance on U.S.-provided networking infrastructure, which may itself engender distrust.

Positive Technical Vectors in Evidence

These considerations are recognized and reflected in both Navy and COCOM initiatives. The committee was impressed by the positive steps being taken to lay the foundation for and to implement information-sharing architectures and coordinated tactical action arrangements with nontraditional partners. More specifically, both the N6-led MDA connectivity effort and the EUCOM-led Gulf of Guinea RMAC Initiative have positive aspects such as the following:

- Putting the keep-it-simple principle into action by sharing unclassified information (e.g., AIS), adopting the commercial Internet model, exploiting COTS products and tools, and so on;
- Leveraging various related efforts, as exemplified by the U.S. emphasis on interagency information sharing in general and on strengthened MDA, to support the NSMS (e.g., the MDA DS COI pilot); and
- Investing in available capabilities to facilitate the exchange of information between the partners ("fly-away" kits, satellite phones, and so on), as illustrated in Figure 3.6.

Additionally, the N6 has developed a multitier (Figure 3.7) graphic as a way to look at the direct sharing of unclassified information with foreign partners while backing this exchange up on the national side of the interface with selected, often sanitized information that derives from more sensitive or classified sources.

This useful DOD-oriented depiction can be generalized as shown in Figure 3.8, emphasizing the creation of a shared information space based on an agreement among partners and supported by partner nations while preserving national information sensitivities. Figure 3.8 reflects the fact, noted by a Chilean Navy officer during discussions with the committee, that all nations have sensitive information content and sources and attendant information sensitivities that must be protected. Accordingly, the reality of information sharing involves some combination of human judgment and prearranged safeguarding technology to filter information in accordance with the range of potentially complex criteria indicated in the figure, including operational sensitivities, capability sensitivities, legal/statutory constraints, and policy/diplomatic constraints. Despite these sensitivities and constraints, it is the committee's view that the resulting shared picture—complemented by trusted person-to-person communication of particularly sensitive information—can provide an adequate basis for cooperative efforts. The JIATF-S operations are evidence of this. Figure 3.8 also depicts an interesting, potentially useful paradigm for collecting and disseminating information that is deemed unclassified: the use of an information broker. The implementation of such a broker concept for MSSIS was discussed above—the accumulation and aggregation of reported AIS information and dissemination of the resulting product to designated users by the Volpe Center.

FIGURE 3.6 Enabling hardware for the user terminal. SOURCE: Paul Dickson, CENTRIXS Operations/Plans, Naval Network Warfare Command (NETWARCOM), "Allied/Coalition CENTRIXS Maritime," presentation to the committee, Washington, D.C., March 13, 2007.

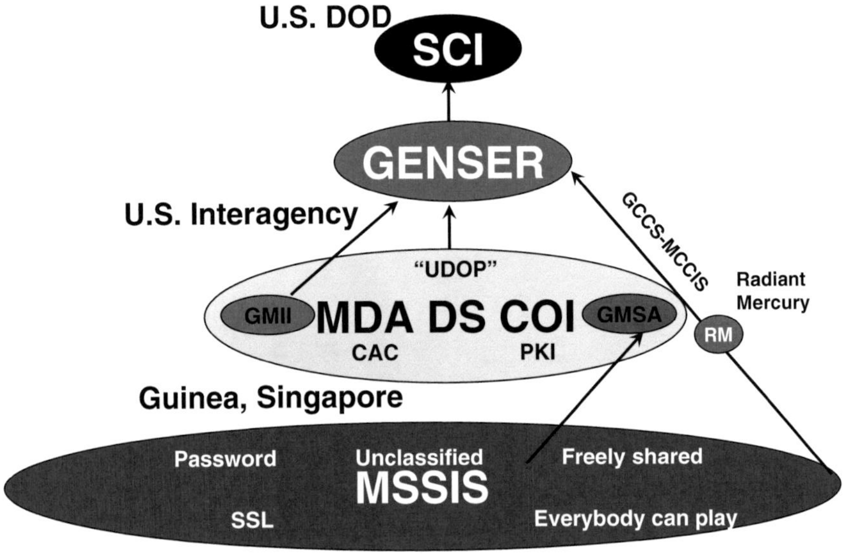

FIGURE 3.7 N6 multilevel architecture depiction. NOTE: MDA DS COI, maritime domain awareness data-sharing community of interest; MSSIS, Maritime Safety and Security Information System; CAC, common access card; UDOP, user-defined operational picture; PKI, public key infrastructure; GCCS, Global Command and Control System; MCCIS, Maritime Command and Control Information System. See also Appendix G. SOURCE: RADM Kenneth Deutsch, USN, Director, Warfare Integration, Office of the Chief of Naval Operations (N6F/N83), "National Academies Naval Studies Board," presentation to the committee, Washington, D.C., January 10, 2007.

In the case of commercial sources and databases (e.g., Lloyd's), this could be accomplished in several ways: by the designation of a national node as an agent for the partnership, through use of a broker for that class of information, or via a commercial node connected directly to the network. The last option, however, could insert commercial players too deeply into the maritime security operations of partner nations.

IT Architectures for Information Sharing

Based on the foregoing observations, the committee offers the following finding and Recommendation 7:

Finding: Effective information-sharing architectures and systems are operating today at the classified and unclassified levels. Navy and combatant commander (COCOM) efforts with nontraditional partners rely on the Internet model and use of commercial products, including for information protection. However, there is

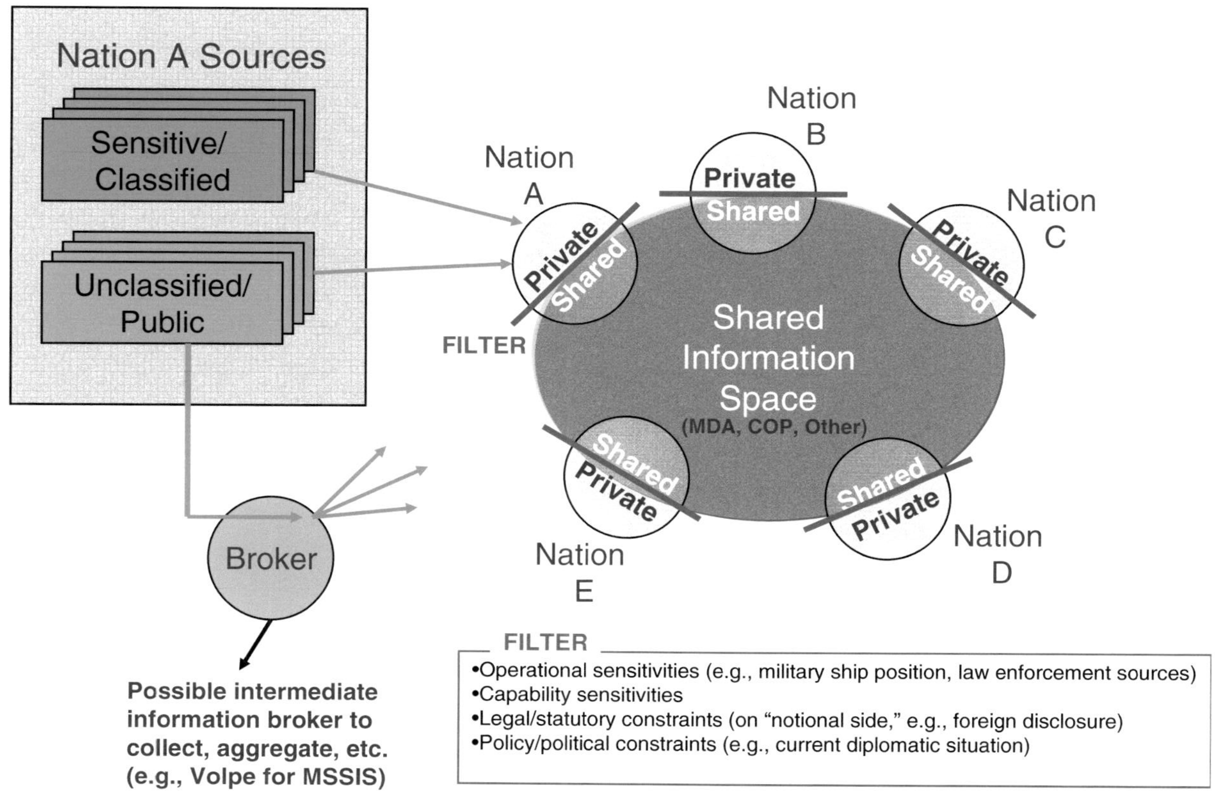

FIGURE 3.8 An operational view of multilateral information sharing.

no known, concerted effort to ensure that the Navy's technical efforts are fully connected to or fully leveraged by COCOM or other initiatives. This less than satisfactory level of effort could lead to interoperability problems or could distract COCOM or other operational elements from their mission focus.

Recommendation 7: The Chief of Naval Operations and the Secretary of the Navy should jointly charter and fund an activity, led by the Deputy Chief of Naval Operations for Communication Networks (N6) and supported by appropriate laboratory/system command/program executive office (PEO) expertise, to provide responsive, dedicated technical support across the full range of interagency initiatives for the design, engineering, and fielding of information technology (IT) infrastructure that would enable information sharing for maritime security.

The activity called for by this recommendation would support combatant commanders, Navy operational elements, other U.S. government organizations, and—through them—foreign partners. It would:

- Develop information-sharing design templates and a catalog of implementing products (these might be different for partners within the U.S. government, those within formal alliances, those in ad hoc coalitions, and those with whom information-sharing arrangements are independent of formal alliance or coalition agreements);
- Assemble and engineer starter kits in support of operational initiatives;
- Include available tools for communications, collaboration, and consultation within the broader design templates, MSP catalogs, and the starter kits effort outlined above;
- Explore potential value-added upgrades for the future and recommend upgrades and backward compatibility approaches;
- Emphasize the sharing of unclassified MDA information, suitably protected to respect privacy and law enforcement concerns; and
- Perform an end-to-end information protection analysis to ensure that the protection meets the expectations of the partners for the several networks in operation or under development.

These measures would increase coherence among inherently distributed regional or subregional initiatives.

Several factors would determine which entity is responsible for providing the technical support called for by the recommendation. These factors would include both (1) the emerging MDA responsibilities and mechanisms outlined above and (2) the critical leveraging of existing organizational capabilities and ongoing efforts that provide similar support. DISA MNIS JPO has a broad charter, along with substantial capability and field presence, albeit focused today on the sharing of releasable Secret information with traditional coalition partners. The SPAWAR System Center in Charleston, South Carolina, has relevant capability and has been supporting, as the committee understands it, the NAVEUR-led Gulf of Guinea JCTD.

The committee understands that its main effort is to identify the job to be done without getting into specific management arrangements. Central to this effort is to specify the content and design of a starter kit along with a set of implementing technologies to support information-sharing initiatives with foreign partners. The discussion up to this point has emphasized the IT infrastructure (e.g., networking). However, as will be noted in subsequent sections, the starter kits should also include (1) available COTS or releasable GOTS tools that provide practical analytical and fusion capability, (2) hardware and software that support operational-level consultation/collaboration and tactical-level action coordination (e.g., satellite phones and chat translators), and (3) commercial information protection tools and technologies.

Regional Information-Sharing Architectures in a Global Context

The committee (1) recognizes significant information sharing today among coalition partners (e.g., CENTRIXS networks and COCOM and Navy initiatives support the multinational coalition in Iraq), (2) then focuses on the adoption of Internet-based IT infrastructures to enable sharing of unclassified information with nontraditional partners, and (3) in Recommendation 7 proposes a Navy effort to strengthen IT-enabling infrastructure architecting, engineering, and fielding in support of MSP initiatives.

As was made clear at the beginning of the chapter, the committee believes that information sharing will generally be carried out among regional or subregional partners, although the global effect will be a collective one. Put differently, the committee does not perceive that MSP success demands global agreements on information-sharing content or top-down enabling system architecture. In fact, it seems clear that the common interests and the requisite relationships are often local. On the other hand, the following also seem to be clear:

- Strengthening the international security regime for information sharing as well as for cooperative action, with the IMO as the central mechanism, should be an objective as MSP efforts proceed.
- The combination of transregional operational situations (such as those that characterize human trafficking) and the potential for CONOPS to evolve as regional MSP matures may lead to more robust information sharing across regions if not around the entire globe.
- Cross-regional information sharing, in turn, calls for an extensible net-centric architecture.

Figure 3.9 attempts to depict the resulting broader architecture—regional information-sharing networks as the core realization of the MSP concept, but

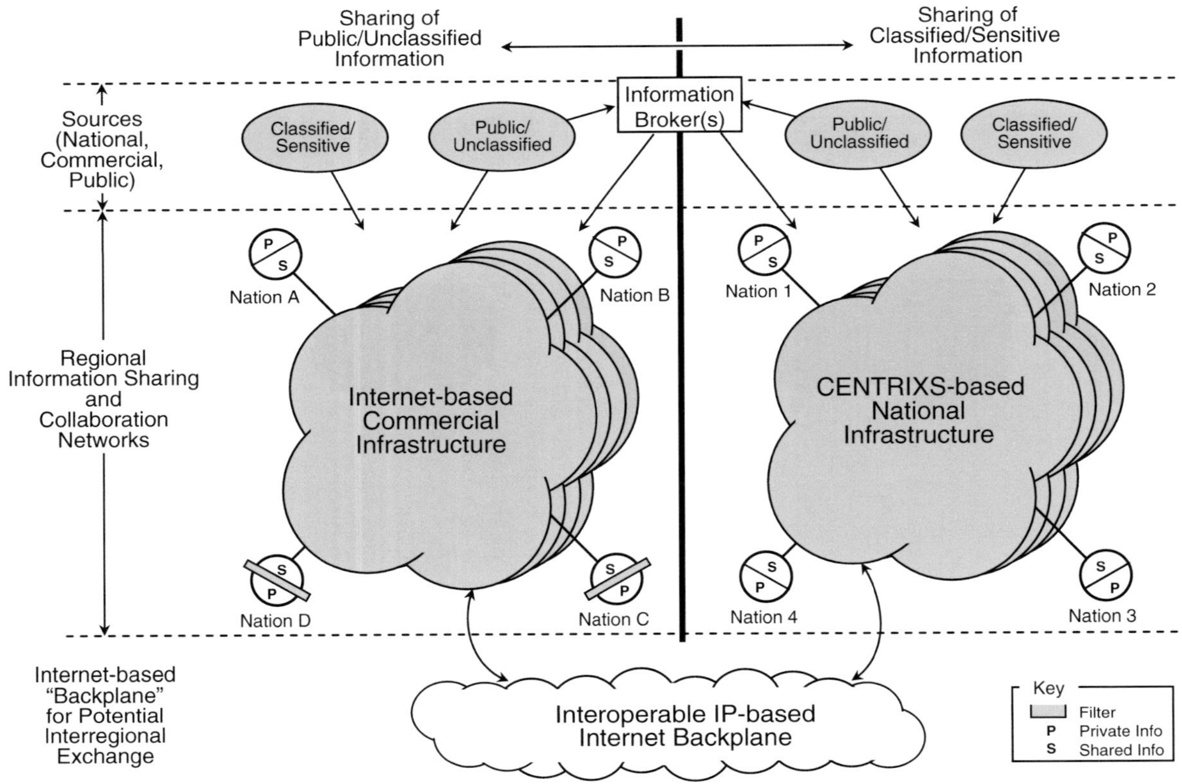

FIGURE 3.9 Enabling maritime security partnerships: a conceptual architecture for information sharing.

with enabling cross-regional networking to support specific operational needs and/or evolving CONOPS.

- The top and middle layers of Figure 3.9, depicting information sources and regional networks, respectively, are generalizations of Figure 3.8. The distinction between networks sharing releasable Secret information and those that share public/unclassified information is represented in a simplified form.
- The lower layer—a cross-regional "backplane"—represents the IP-, Internet-based networking capability offered by modern global technology. It provides the potential to reach beyond regions as the international maritime security community matures and evolves.

The architecture depicted in Figure 3.9 provides for both interoperability (IP-based networking, common data vocabulary/representations within the COI, and the like) and commonality of technology/product building blocks (when doing so makes sense). Again, Navy efforts to pursue such an architecture are noted and applauded.

The point here is not that information sharing within and beyond regions is easy. There are the challenges of achieving agreements that are actionable, protecting the legitimate information sources of partner nations and other information providers (e.g., commercial shippers), and so on, as discussed throughout this report. Rather, the point is that an extensible information system/networking architecture based predominantly on modern commercial technology can be practically envisioned, which enables rather than limits progress toward MSP objectives.

BUILDING MISSION CAPABILITY

The discussions above recognize and applaud the ongoing initiatives and efforts to advance information sharing and coordinated action capabilities with and among nontraditional partners. Much of the Navy and COCOM-related material discussed above dealt with architecting and prototyping early instantiations of an enabling net-centric information infrastructure and implementing useful, basic AIS-oriented information sharing.

However, if the objectives of the MSP initiative are to be realized, it seems crucial to move beyond the enabling information infrastructure and the sharing of readily available, nonsensitive, unclassified information. More effort is needed to strengthen the mission capability that would employ this infrastructure: focusing on information content; enriching information sources and their coverage; enhancing information analysis/data fusion capabilities; and exploiting a rich menu of available tools to support collaboration and coordination. And, in all of these areas, it is important to investigate the role of advancing technology while recognizing that the challenges confronting new and emerging partnerships reside first with policy and trust, not with technology.

The systematic exploration of capability enhancements requires some scheme for narrowing down the requisite functional building blocks and identifying the interactions and trade-offs among these. A generic security engagement chain based on an "interdiction continuum" presented to the committee by a JIATF-S representative is one such scheme (Figure 3.10). The mission execution process involves continuous feedback among the elements of the chain.

For the purposes of this report, the committee adopted a three-part, somewhat simplified functional breakdown of the elements of the chain or, more broadly, of the classical C2 process:

- *Sense/collect.* Defined to include partner as well as U.S. capabilities, ranging from technical surveillance (e.g., radars), through automated electronic reporting (e.g., AIS), to human reporting (e.g., local law enforcement);
- *Analyze/fuse.* Defined to include the exploitation of multiple sources to improve the quality of a single class of information (e.g., vessel tracks) and to derive broader information (e.g., connecting multisource dots to detect suspicious patterns); and
- *Decide/act.* Defined to include effective mechanisms, including feedback/monitoring for coordination and consultation during the decision process and for the exercise of C2 once a decision has been made.

Sense/Collect

The effectiveness of any maritime security information-sharing regime will ultimately be limited by the quality, completeness, and timeliness of the underlying information sources. As shown in Figure 3.11, two broad sources of information relevant to improving regional and global maritime security can be identified—intelligence and surveillance:

- *Intelligence.* Traditional and nontraditional reporting of a broad spectrum of information relevant to maritime security includes clandestine human intelligence collection and reporting to overtly accessing a wide range of commercial and law-enforcement-related sensitive but unclassified data. Intelligence is defined here to include the information content of intercepted communications, as distinct from the possible surveillance value (location) of such intercepts. It also includes information gleaned from ship boardings. Intelligence is generally not very close to real time, so surveillance assets must often be employed to find a vessel of interest on the high seas and to take tactical action.
- *Surveillance.* Reporting from all sensors on detection, identification, and tracking of ships and craft on the surface of the ocean. Includes a broad spectrum of sources, ranging from coastal vessel detection and tracking radar systems

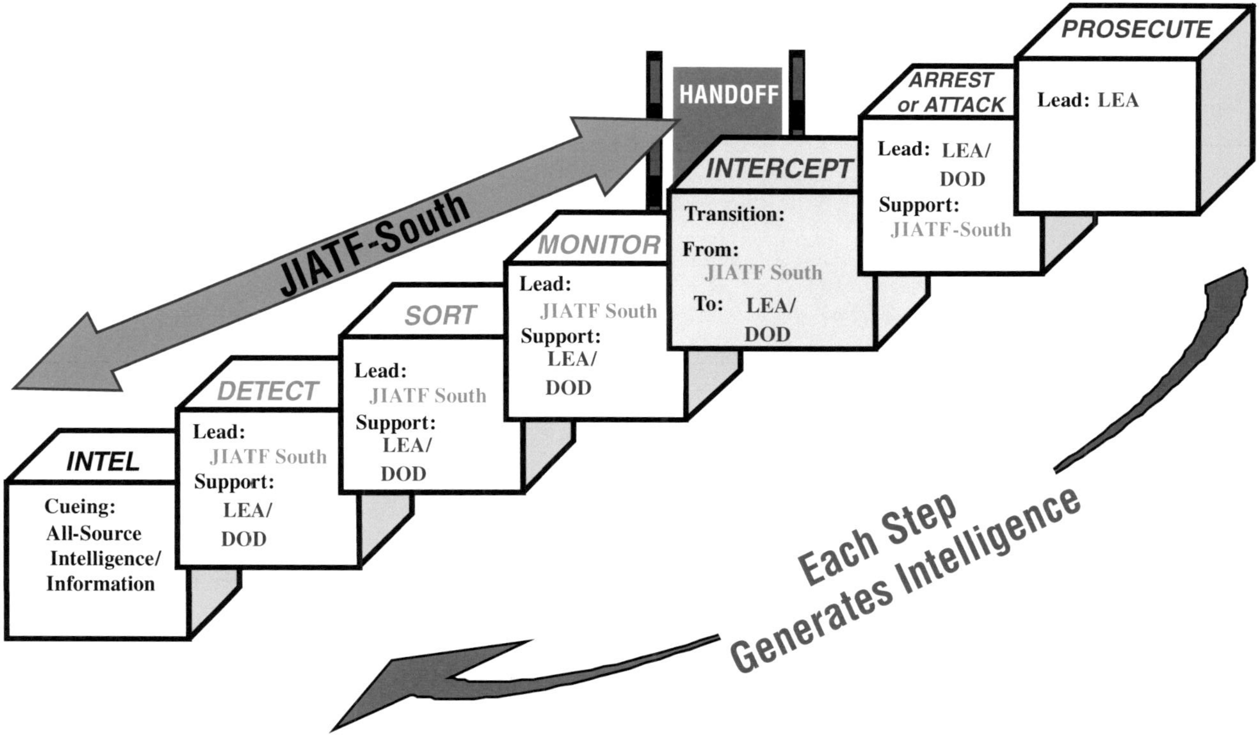

FIGURE 3.10 A representative maritime security engagement chain. NOTE: LEA, law enforcement agencies. SOURCE: Scott Cantfil, Joint Interagency Task Force-South (JIATF-S) liaison officer, "Interdiction Continuum," presentation to the committee, Washington, D.C., May 16, 2007.

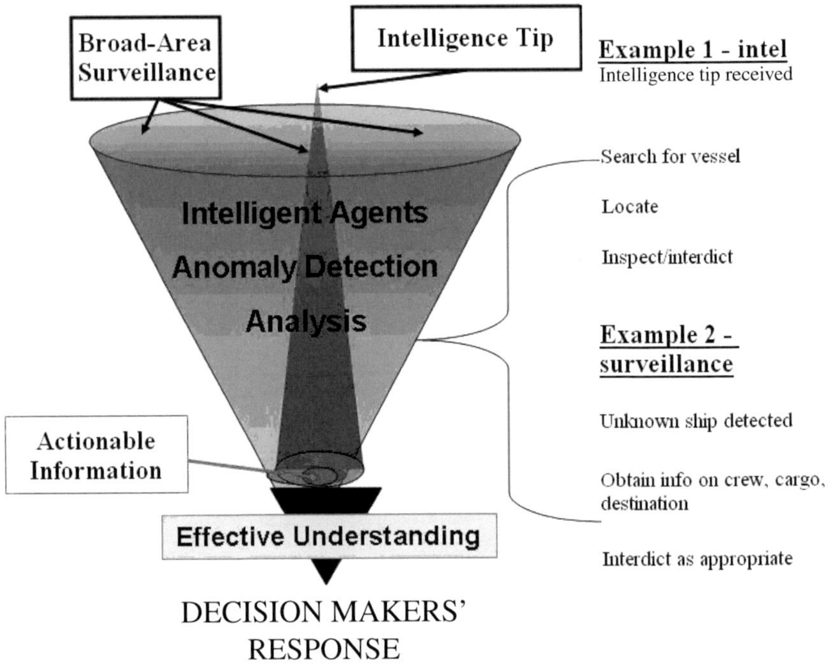

FIGURE 3.11 Two sources of information: broad-area surveillance and intelligence tips. SOURCE: Based on RDML Joseph L. Nimmich, USCG, Assistant Commandant for Policy and Planning, COMMANDER, USCG SECTOR KEY WEST, presentation to the committee, Washington, D.C., January 9, 2007.

to highly classified U.S. "national technical means." (The notion of surveillance, broadly defined, of course applies to people and cargo as well as ships. However, the surveillance of people and cargo results in what is here defined as "intelligence.")

While the distinction between intelligence and surveillance sources is important, it is also important to recognize that the two mechanisms are frequently interdependent in the operational arena, such as when surveillance assets are required to detect and track a vessel that has been previously reported by intelligence to be of interest to maritime security officials.

Intelligence

The United States and other nations concerned with the many facets of maritime security, ranging from the efforts of the Proliferation Security Initiative

(PSI) to reduce the threat of weapons of mass destruction (WMD) smuggling to the routine interdiction of migrants, are now primarily dependent on intelligence cueing for initiating operational responses. Much of this intelligence is from law enforcement sources.

The United States has had considerable experience in the collection and use of intelligence/law enforcement information in pursuit of its maritime security goals. U.S. expertise is currently located at two operational centers, the JIATF-S and the National Maritime Intelligence Center (NMIC).

The organizations charged with preventing the smuggling of drugs into the United States by sea, primarily from South and Central America, rely on a robust information network, CNIES, to provide information on planned and ongoing maritime drug-smuggling activities. The nature and quantity of these inputs tax the ability of the available response and interdiction forces to take advantage of all such tips. JIATF-S is unique in having established effective procedures for routinely and quickly converting classified intelligence and sensitive law enforcement information into a form that can be shared at an unclassified level under bilateral agreements with partner nations capable of taking responsive actions. The CNIES information-sharing system is described in the section "Systems for Sharing Information."

JIATF-S experience in the Caribbean, as understood by the committee, is that broad-area, uncued maritime surveillance is a less important source of information about seagoing drug traffic than is intelligence. However, a considerably more local, directed surveillance effort is often mounted by response forces to convert intelligence tips into actionable ship tracks and identities. It would appear that improved broad-area maritime surveillance (BAMS) in the Caribbean and Eastern Pacific, for instance, could significantly improve the efficiency of intercept and interdiction assets. Understanding that achieving such surveillance capability presents its own challenges, the committee is unaware of any existing operations analysis that would help identify the most cost-effective combination of surveillance and intercept capability, although JIATF-S headquarters is beginning to collect the data that would be essential to such analyses.

The second focus of information collection and use, the NMIC, is now under the purview of both the Director of Naval Intelligence and the USCG. This major operational intelligence facility has been expanded to keep track not only of all information on shipping worldwide (including warships) as traditionally reported by intelligence and reconnaissance sources, but also of information on the emerging maritime security challenges of particular concern to the United States. NMIC's primary purpose is to provide timely maritime intelligence support to Navy and USCG elements and to other government agencies needing such information. The NMIC has well-established links with traditional U.S. allies, and, in addition to the information it obtains from highly classified intelligence collection systems, it leverages law enforcement information, as appropriate, and increasingly takes advantage of a broad range of commercially available maritime information. However, because the NMIC's primary sources are so sensitive, it is difficult to establish procedures for transferring information in a timely way to the nontraditional partners that are the focus of this study. Thus the information-sharing architecture postulated by the Navy and generalized in this chapter involves filters that reside between national sources (foreign as well as U.S.) and the shared information picture or database (see Figures 3.7 and 3.8), and the level of information to be shared may differ among the four levels of partners mentioned in the terms of reference.

Vessels of Interest A vessel might be designated "a vessel of interest" to one or more agencies responsible for some aspect of maritime security depending on the information available on its history, crew, cargo, and movements. Such information exists in a variety of forms and locations, ranging from shipping documents and including the content of automated reporting systems such as AIS as well as specific tips from intercepted communications or direct observation reported by the intelligence and law enforcement agencies of one or more cooperating nations (e.g., customs analysis of ship manifests). Such information might be sufficient in and of itself to arouse interest in a specific ship; alternatively, disparities in the information about a specific ship provided by different sources might prompt further investigation.

Implications for Information Sharing As noted above, sources of intelligence on maritime traffic range from traditional highly classified national intelligence collection and reporting systems that were originally developed primarily to deal with military ships, through a rich set of law enforcement information, to the broad and increasingly important category "commercial and nontraditional." Information from all these sources, when fused with complementary data from a surveillance system's sensors, could, in principle, provide a comprehensive COP of all activity on the surface of the ocean, or at least of all activity in an area of interest.

In addition to the various human and technical information collection practices of nations, many of which reside within the various national security and military organizations, the various law enforcement communities in most nations maintain databases on individuals and vehicles that fall into the general category "law-enforcement sensitive." These include watch lists of known or suspected terrorists and long-standing Interpol procedures for exchanging information on specific individuals who are formal subjects of arrest or detention warrants. Similar information collection and sharing arrangements exist and are being expanded by the customs officials of several coastal countries.

The growing exchange of law-enforcement-sensitive information in support of the MSP initiative could be very productive but is fraught with privacy and legal issues. For example, the United States and the European Union have engaged in a long-running dispute about the level of detail needed in the information to be exchanged concerning passengers on flights bound for the United States. In any case, it is clear that increased sharing of these types of intelligence data would make maritime activities more transparent—a central theme of the MSP initiative.

Nontraditional Sources of Intelligence In addition to governments (U.S. and non-U.S.) as sources of maritime information, there are rich sources of intelligence that are not part of a nation's formal intelligence and law enforcement collection and reporting systems.[12] Increasing reliance on nontraditional sources of maritime information will greatly enrich the MSP information-sharing concept, and the concept will face fewer bureaucratic and political challenges.

Examples of nontraditional sources of maritime intelligence information that could contribute to information sharing among MSP include the following:

- The Global Integrated Shipping Information System (GISIS) (operational, under IMO, Web-based),
- International LRIT Data Centre and Data Exchange (in development by IMO),
- Port state information exchanges
  —Equasis (operational, Web-based)
  —European Communities: SafeSeaNet (operational, Web-based)
- Information on fishing vessels: Fisheries Global Information System (FIGIS) (satellite/VMS-based; being fielded by the Food and Agricultural Organization), and
- International Network for Cooperation and Coordination of Fisheries-Related Monitoring, Control and Surveillance (MCS) activities (operational, under the National Oceanic and Atmospheric Administration).

Summary The proper use of intelligence information for either establishing a COP or identifying a particular ship as "of interest" is very complex, and many protocols have yet to be worked out. Much of the intelligence about ship crews and cargoes is law-enforcement-sensitive and subject to disclosure rules that can be even more restrictive than national security classification protocols. In addition, intelligence systems that access proprietary commercial databases are also highly circumscribed in their ability to share information widely. Nevertheless, a fully effective, shared maritime security information system will need to integrate as much intelligence data as possible.

Sensitivities and constraints can often be handled by sharing only the operationally significant "finding" (e.g., a tactical alert without any trace of sources or

---

[12]Appendixes C and F discuss international databases as potential sources of such shared information.

methods). This paradigm can support the operational mission while protecting the legitimate information concerns of partner nations.

Observations on Intelligence Despite the apparent success of intelligence cueing as a primary source of information that enables maritime security operations, there is growing concern about the potential threats posed by vessels that are not typically subjected to the type of information reporting currently available. Such vessels are generally smaller than the vessels routinely reported on in intelligence and commercial shipping circles (except for drug boats), and they operate from noncommercial ports, where they are less visible than they would be in highly regulated commercial ports. Pirate ships and human traffickers are examples of such threats, if not directly to U.S. interests then to the interests of many of the nontraditional partners to which the United States seeks to provide useful information in order to gain better maritime cooperation. Future threats could well include the smuggling of WMD or direct attacks by low-visibility, noncommercial vessels.

More broadly, there is a large class of ocean-going ships and craft that are not routinely subjected to observation and reporting. These are the myriad unregulated private craft and other vessels that can, if they choose, generally remain unobserved by existing reporting and surveillance systems that depend on radio frequency emissions. Such vessels are acknowledged to constitute a growing source of threats to maritime security from the smuggling of contraband and people, the poaching of resources, piracy, and even attacks from the sea.

It is highly doubtful that even planned improvements in traditional intelligence and law enforcement collection and reporting capabilities will provide enough information about the existence and location of such threats to preclude their deployment. Detection of ships of this type will require better broad ocean surveillance capabilities than are available today.

Surveillance

As noted in the foregoing section, intelligence currently provides much of the information that enables responses to threats to maritime security. This is not to suggest that routine, uncued surface surveillance is currently unimportant or unproductive.

Nations concerned about the enforcement of laws designed to protect marine fisheries and other resources frequently conduct routine or randomized surveillance patrols of high-value areas employing both ships and aircraft. Known smuggler's routes are also routinely patrolled by ships and/or aircraft to deter and interdict if needed. This includes, for example, the USCG patrols in the Mona and Windward Passages and the Strait of Florida and the recently expanded cooperative patrolling and coastal radar surveillance of the Strait of Malacca by the riparian countries. Other examples include the routine employment by the

United States of its fixed-site over-the-horizon radars (OTHRs) to detect and track aircraft in the Caribbean and the use by many nations of coastal surveillance radars.

Some nations also operate surveillance satellite systems that serve multiple purposes,[13] including the detection and location of ships at sea—primarily those transmitting on various radio frequencies (mainly surface search radar frequencies). Such detection systems help to populate geographical plots (such as those at the NMIC) with large numbers of ship locations and, in some cases, identities. However, this may do little to enhance maritime security unless mechanisms exist for identifying specific ships as being "of interest" on the basis of supplemental intelligence information, as discussed above, or on the basis of anomalous behavior, as discussed below in the section "Analyze/Fuse."

The fact that the effectiveness of existing, mostly passive, broad ocean surveillance systems is highly dependent on the "cooperative" radiation of ships indicates that complementary active systems—primarily radar systems—are thought to be needed if truly persistent surveillance of important broad ocean areas is to be established, as envisioned by the U.S. NSMS and the associated MSP concept.

U.S. Capabilities for Active Maritime Surveillance The USCG is currently expanding its radar surveillance of port and harbor approaches, but such efforts are inherently restricted to relatively short inshore ranges. At present the U.S. ability to actively surveil broad ocean areas is concentrated in its fleets of maritime patrol aircraft, operated principally by the Navy but also by the USCG and the customs and border patrols. The Navy's capabilities are focused on protecting its forward-deployed military task forces against potential threats. The protection capabilities of the Navy's maritime patrol aircraft fleet are planned to be modernized—note, in particular, the current competition for a force of long-range, unmanned aerial vehicles for BAMS which can sustain five orbits when fully fielded. Little naval surveillance capacity is expected to be allocated for routine active surveillance in support of national or regional maritime security objectives other than fleet protection. One exception to this general observation is the ongoing surveillance support to maritime interdiction operations (MIOs) in the Persian Gulf. These operations help to enforce shipping laws and regulations and prevent the seaborne introduction of contraband into the Iraq theater of military operations.

To date, the Navy has not found the need for more effective BAMS to be sufficiently compelling to initiate a major new sensor acquisition program for this purpose, such as a space-based radar, a fleet of high-altitude airships, or any one of a number of other expensive schemes that have been identified. Any such new

---

[13] See <http://www.centennialofflight.gov/essay/SPACEFLIGHT/recon/SP38.htm>. Accessed August 28, 2007.

program might come at the expense of other DOD—probably Navy—programs now deemed of more immediate importance.

Instead of focusing on new or improved sensors that would extend the nation's surveillance capability to include uncooperative ships (as detailed in the section "Current and Emerging Information Architectures"), the Navy has chosen to focus its near-term efforts on making better use of the intelligence and passive surveillance information that is already available through improved fusion and analysis. There is considerable merit in these activities in that, when fully implemented, they should permit both greater responsiveness to intelligence cueing and, if it proves to be useful, the employment of anomaly detection concepts that could spotlight potential problem ships that are trying to appear legitimate by hiding in plain sight (see section "Surveillance," below). Such anomaly detection concepts offer the only known way of identifying apparently compliant ships as potential problems if no tip is available—short of a dramatic and unlikely upgrade to the regulations for ships at sea that would be comparable to the regulations for aircraft operators.

Potentially Affordable Surveillance Improvements Worth Additional Attention

Historically the Navy has been the nation's center of excellence for ocean surveillance, having fielded many innovative concepts over the years, from the airships used for coastal surveillance patrols in World War I and World War II, to the Cold War's acoustic and electronic intelligence (ELINT) satellite systems.[14]

As suggested above, it appears to the committee that the Navy has taken an understandably cautious approach to expanding its maritime surveillance capabilities to meet the surveillance challenges of the National Strategy for Maritime Security. While caution is clearly called for before the Navy becomes committed to a major new acquisition program and its attendant future operating costs, the committee believes that there are several potentially less costly opportunities for improved ocean surveillance that warrant technical development and concept exploration. It also believes that the U.S. Navy is in the best position both technically and in terms of its traditional role and mission to pursue such opportunities. The next seven sections highlight those concepts that have elicited the greatest interest on the part of the committee.

Exploitation of Data from Commercial Remote Sensors/Satellites. The committee was impressed with the potential for fusion of data streams from current and planned commercial imaging satellites, both electro-optical and radar. A demonstration was witnessed at the University of Miami's Center for Southeastern Tropical Advanced Remote Sensing (CSTARS) facility. While rudimentary, it indicated considerable potential for the employment of such data streams in

[14]See <http://www.centennialofflight.gov/essay/SPACEFLIGHT/recon/SP38.htm>. Accessed August 28, 2007.

tracking uncooperative ships. At the time of the CSTARS experiment, feeds from only eight satellites were available. As shown in Figure 3.12 and in more detail in Appendix E, by the end of this decade there will be 58 commercial optical satellites on orbit, up from 31 now, and 13 radar satellites, up from 4 now (all foreign).

Open ocean surveillance using commercial space-based imaging appears to have considerable potential. The committee performed a notional analysis of space-based imagers to assess the potential surveillance regions. It assumed a 50-nmi radius imaging footprint and a nominal resolution of 1 meter for the purpose of reidentification during each orbital pass. This resolution is consistent with 42 percent of the satellites shown in Figure 3.12. The resulting imaging footprint is larger than the individual footprints of these very high resolution systems and would represent the total footprints of several such satellites. Assuming a nominal 1,000-nmi orbital altitude, such an imaging satellite could sweep out 1.1 million sq nmi per hour. Of course the orbital trajectory would limit the field of view of such a vehicle. Another consideration is the cost of the large amount of data that would be needed if the satellite were to provide enough images for large-area coverage.

For a 100 nmi by 100 nmi swath with 1-meter resolution (again, representing four or more satellites), data reporting and processing rates would probably exceed gigabits per second. If the satellites transmitted imagery only when a detection was made, the data rate would be much more manageable. For example, assuming that the data content of the message, including imagery, is about 1 million bits, then for an open ocean density of one ship per 1,000 sq nmi, only 3.3 Mbps of downlink data would notionally be required. For denser environments, this rate would increase considerably. The committee concludes that if automated features were in place for reporting on only detected vessels and with cued imaging of limited areas of ocean, satellite downlink data rates should be well within existing capacities.

For such a scheme to be implemented, the satellite design would need to include appropriate onboard software to automate the detections-only concept, and the nations interested in using such satellites for ocean surveillance in this way would need to come together and perhaps subsidize the development of such features.

Potential Upgrade of U.S. Over-the-Horizon Radars (OTHRs). The United States has a long history of employing continental United States (CONUS) fixed-site low-frequency OTHRs for the long-range detection of aircraft that may be threatening the United States. However, the use of such bistatic-Doppler radars for detecting and tracking surface ships and craft is impeded by the need to deal with the high level of clutter returned from sea waves.

The U.S. Navy has taken the lead on behalf of the DOD in responding to a Presidential Directive for DOD to improve the detection and interdiction of drug

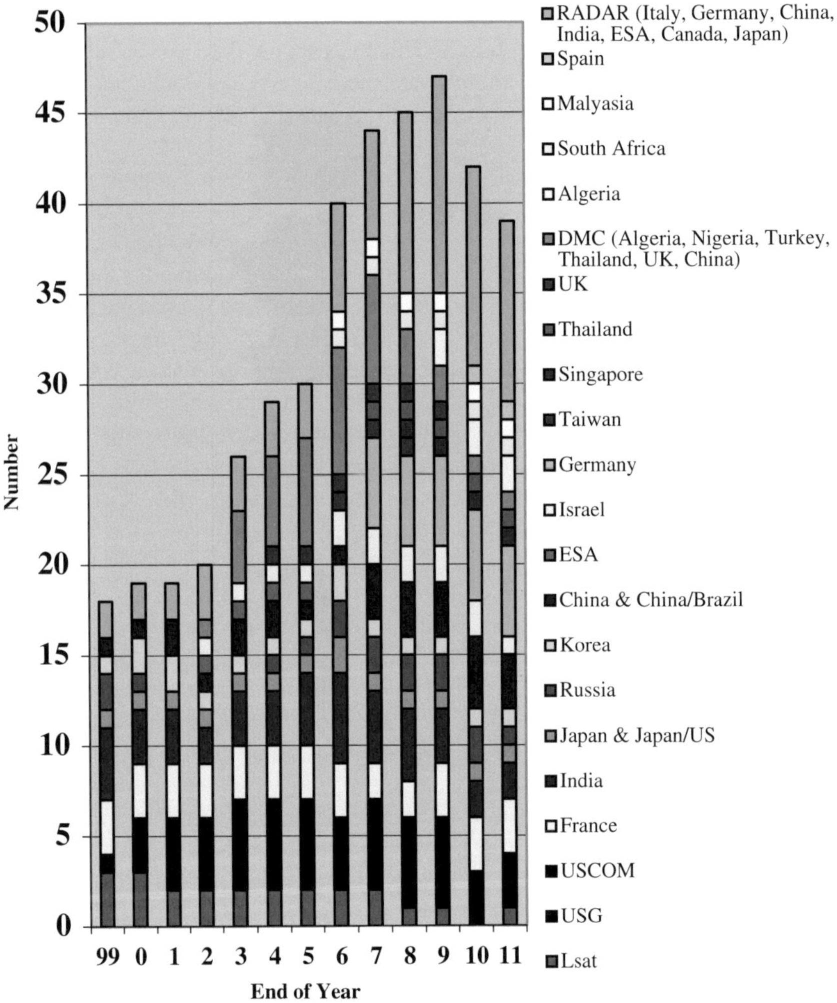

FIGURE 3.12 Number of optical and radar land-imaging satellites. SOURCE: William E. Stoney, Mitretek Systems. 2006. "ASPRS Guide to Land Imaging Satellites," updated for the NOAA Commercial Remote Sensing Symposium, Washington, D.C., September 12-14. Noblis, Inc. ©2007. Reprinted with permission.

traffickers en route to the United States. The Navy retained three relocatable OTHR (ROTHR) sites (Texas, Puerto Rico, and Virginia) after the end of the Cold War and now operates them in support of the JIATF-S drug interdiction mission, as described below. To that end, there have been only modest improvements to these systems over the years that have helped in the detection of surface craft, in addition to the main goal of detecting light aircraft (see Figure 3.13 for current coverage of the aircraft).

The committee did not assess the degree to which further improvements to these obsolescent radars targeted on surface vessels of a particular size, for instance, would be more cost-effective than their replacement with modern equipment. However, the prime contractor for the Navy's ROTHR program, Raytheon, has proposed an expansion of the ROTHR coverage primarily in support of U.S. homeland defense/security, as shown in Figure 3.14. This suggests that a body of technical expertise is well established and available to the Navy and could be offered to other countries in support of the overall goal of improving the persistent ocean surface surveillance capability worldwide as a key to a successful MSP initiative.

The Air Force OTH-backscatter radars (FPS-118), whose coverage is shown in Figure 3.15, became operational near the end of the Cold War and are now in warm storage, although they have undergone some testing by NOAA for observing ocean surface parameters. Because these radars are owned and managed by the Air Force, whose mission does not normally include surveillance of ocean surface traffic, there appears to be little information on their prospective utility for

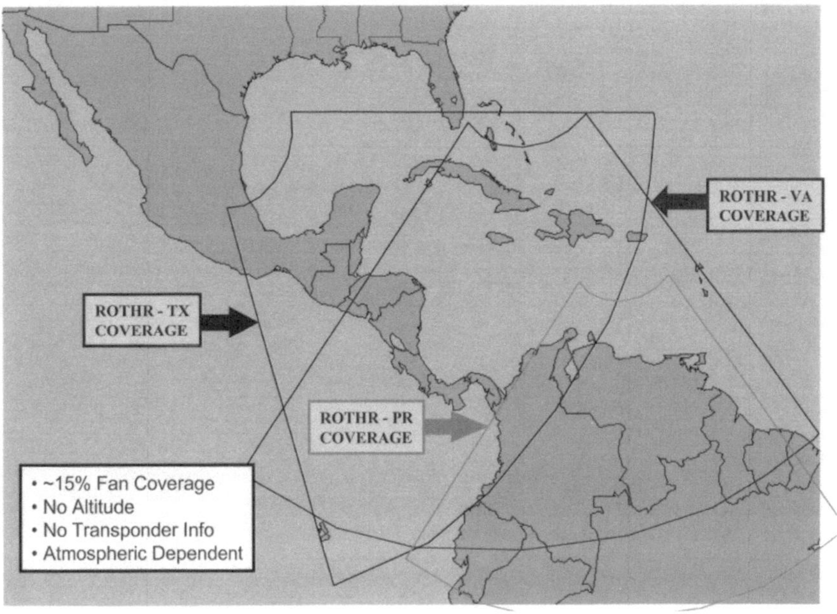

FIGURE 3.13 Existing relocatable over-the-horizon radar (ROTHR) coverage (JIATF-S).
SOURCE: Joint Interagency Task Force-South, "ROTHR Coverage," June 25, 2007.

ocean surface surveillance. However, they may have potential for ocean surface surveillance, which could be explored.

Because OTH systems might be able to provide information on surface traffic far off each coast of the United States, as well as near other countries, analysis of the cost and effectiveness of hardware modernization and improved signal processing would appear to be warranted. Because the Navy is the de facto U.S. leader in ocean surface surveillance, Navy leadership of such an exploratory process would be essential.

Potential for Improved Regional Maritime Surveillance Through Expanded Coastal Radar Surveillance Systems and AIS Receivers. The committee has been impressed with the efforts of the naval component commanders working under the regional COCOMs to encourage and assist other countries in improving their maritime surveillance capabilities. EUCOM is devoting considerable effort to assisting coastal nations throughout its theater, including particularly the Gulf

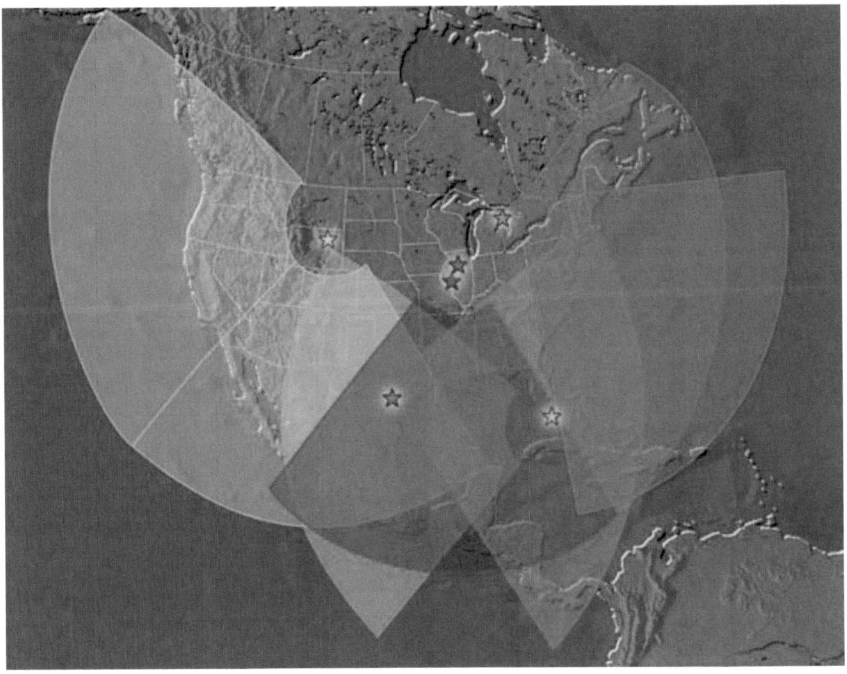

FIGURE 3.14 Prospective ROTHR coverage (Raytheon). SOURCE: Reproduced with permission from Raytheon Company, ©2004, relocatable over-the-horizon radar (ROTHR) for homeland security. See <http://www.raytheon.com/products/stellent/groups/public/documents/legacy_site/cms01_049201.psf>. Accessed August 28, 2007.

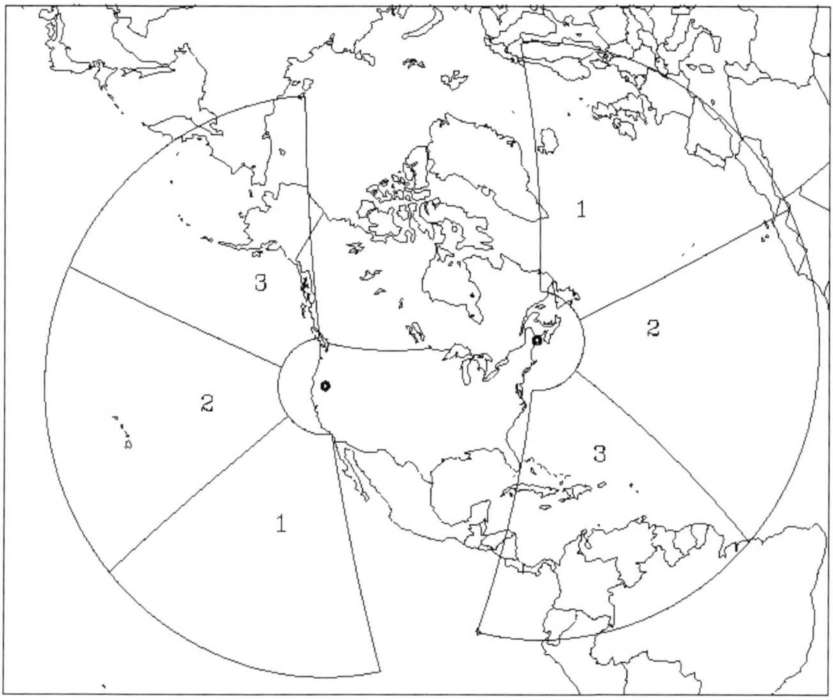

FIGURE 3.15 OTH-B. SOURCE: Federation of American Scientists. 1999. U.S. Air Force, over-the-horizon-backscatter (OTH-B) air defense radar system, June 29. See <http://www.fas.org/nuke/guide/usa/airdef/an-fps-118.htm>. Accessed August 28, 2007.

of Guinea, to improve their maritime surveillance capabilities by installing coastal AIS receivers and radars and to provide the resulting information to the MSSIS Web-based distribution system. PACOM has been quite successful in encouraging cooperation in maritime surveillance and information sharing among the nations of Southeast Asia, particularly in the vicinity of the Strait of Malacca.

Given these encouraging activities, it appears that a Navy-led initiative to expand such efforts worldwide and to include the cost of such technical assistance efforts in the Navy's baseline budgets, rather than relying on the ephemeral nature of most COCOM direct funding sources, could have a significant payoff. To the extent that U.S. Navy expenditures for the technical assistance that other countries need to become full partners in the provision of ocean surveillance information are modest, such an initiative might be very cost-effective.

Taking Advantage of Existing Commercial Ship Surface Search Radar and AIS Data. In addition to the dedicated active and passive ocean surveillance systems that are already designed to feed various monitoring, analysis, and control centers, most ships and aircraft at sea, military and civilian, operate their own radars in the interest of safe navigation and the avoidance of collisions. It has been suggested that the local information from such sensors could greatly expand the coverage of networked ocean surveillance sensors.

Of particular interest to the committee is the potential harnessing of the local radar and AIS displays available on all commercial ships that are already subject to IMO agreements. Given the tens of thousands of such ships that are usually at sea, this appears to constitute a significant source of surveillance information if it can be tapped at reasonable cost. The U.S. Navy is already fielding a SureTrak capability for its ships to integrate its own ship radar and AIS receiver data. Presumably that integrated picture can be transmitted ashore over military communications channels and integrated into the expanding U.S. MDA information system. The extent to which such information derived from military sources could be routinely shared with prospective MSP nations should be investigated. It should be noted here that the number of commercial ships at sea is much larger than the number of military ships and could considerably expand coverage.

Figure 3.16 indicates the relative historical densities of several types of ships per square degree of latitude and longitude, including fishing vessels, merchants, and tankers. Smaller ships are not included. The maximum density (lightest shade) is greater than 25 ships per square degree. At about 50 degrees latitude,

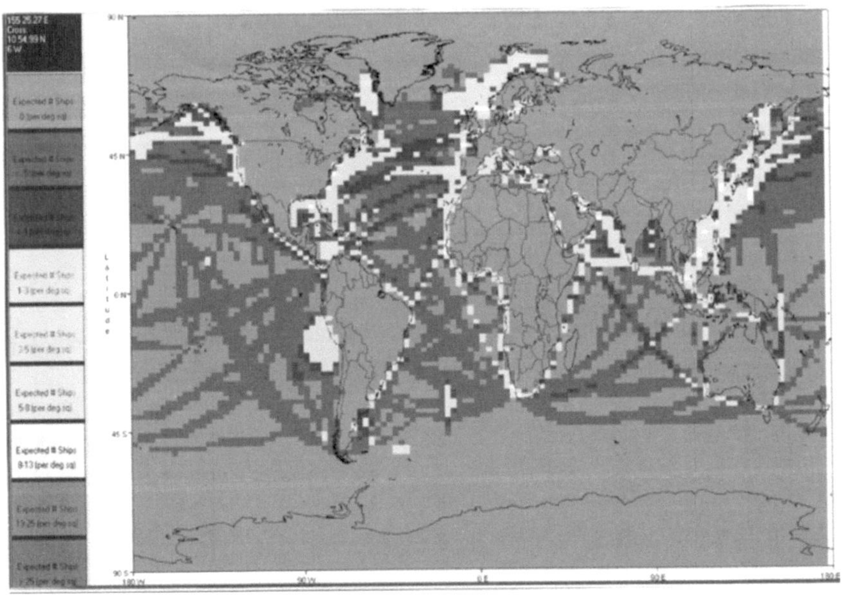

FIGURE 3.16 Commercial shipping density.

this corresponds to more than one ship per 110 sq nmi. As a check, the committee examined an AIS snapshot near Portsmouth, England, taken on October 8, 2004, at 13:03:28 UTC, the area of highest density in the figure. The committee counted 43 ships per square degree, or one ship per 64 sq mi, which is consistent.

As shown in Figure 3.16, the average density of commercial ships at sea is highest in the approaches to commercial harbors, along coastal routes, and in and near well-known choke points, as would be expected. This suggests that commercial shipborne navigation radars offer the potential for sufficiently robust coverage in many areas of the world to largely preclude any ship from being able to slip through undetected and reported. Even in less frequented areas, ships attempting to remain undetected would be greatly challenged to find an evasion route.

For such a concept to be effective, there would have to be an affordable method of piping the available surveillance information ashore from operationally significant standoff distances. The potential cost to shipowners of using commercial satellite telephony for such purposes appears to have inhibited serious exploration of this concept to date. Fortunately, modern image/data compression techniques promise to reduce the file size of such periodic reports to manageable levels. Such techniques, when combined with emerging government-sponsored (and perhaps government-subsidized) satellite communications services such as those being provided by Increment 3 of the DHS/USCG NAIS program and other systems that utilize the Iridium, ORBCOMM, or GlobalStar satellite communications constellations, offer considerable promise for providing the needed linkage at acceptable cost.

For example, a simplified analysis indicates that in an area from which 1,000 ships are reporting every ship contact from radar or AIS with a 1,000-bit message every 15 to 60 minutes, a ship would require a data rate of no more than about 11 kilobits per second (kbps) even in areas of extremely dense traffic (such as one vessel per square nautical mile). At current International Maritime Satellite (INMARSAT-C) rates, this type of reporting incurs a modest price and is well within the capacities of commercial satellite communications systems. Assuming a radar coverage range of 24 nmi and an AIS range of 50 nmi, 1,000 ships with no overlapping coverage could cover 8 million sq nmi for AIS and 2 million sq nmi for radar. At an average ship speed of 15 kt (based on a cursory examination of vessels described in *Jane's Merchant Ships*, June 2007), 1,000 ships would sweep out about 1.5 million sq nmi per hour for AIS and 0.75 million sq nmi per hour for radar, presuming that coverage separation remained the same. An isotropic distribution of ships is of course not likely, so the foregoing estimates are upper limits, but real-world performance of even half that amount could be very significant. It might, for example, provide nearly solid coverage of the U.S. East Coast out to 200 miles.

As with other sources of information on commercial ship location, cargoes, and routing that are potentially competition-sensitive, the data handling system would have to provide appropriate safeguards.

Feeding Routine Offshore Surface Contact Data Generated by Military Ships and Aircraft into the Surface Ship Database. As noted above, the Navy is already planning to outfit at least some of its ships with the SureTrak feature that integrates surface search radar and AIS receiver information. A more general and widespread implementation of this concept among all of the navies of the MSP nations could further increase active surveillance coverage. If cooperating nations were to further expand the concept to include all of their radar-equipped government ships and aircraft, additional coverage would become available. Again, the U.S. Navy is well positioned to provide technical assistance for such an MSP initiative. A first step would be to seriously analyze the potential for coverage and its utility, along with cost.

Reactivation and/or Expansion of an Integrated Undersea Acoustic Surveillance System. The U.S. Navy has a long history of successfully employing fixed and mobile passive acoustic surveillance systems to detect vessels of interest in key areas of the world. During the Cold War, those vessels of interest were almost always Soviet, particularly the submarines and other warships. These surveillance systems were generally considered "fleet assets" and were managed outside the purview of the intelligence community—an arrangement that has certain "optical" benefits in the current international environment.

With the end of the Cold War, the need for such labor-intensive acoustic surveillance systems declined appreciably, but the relevant expertise has been retained and even extended in the form of experimental distributed underwater arrays, towed surveillance arrays, tactical sonar systems, advanced sonar buoys, and acoustic capabilities for unmanned underwater vehicles. As with the airborne surveillance provided by MPA and BAMS, these acoustic surveillance efforts are currently focused largely on fleet protection, not persistent, broad ocean surveillance of the type envisioned by the MPS initiative.

Modern signal processing techniques and automated data handling suggest that increased utilization of acoustic surveillance concepts may complement the other enhanced surveillance concepts summarized above. As with specific emitter identification (SEI) techniques employed in the passive ELINT regime, the acoustic signatures of ships can provide considerable useful information in addition to the location of the ship. As the nation's expert in ocean acoustic surveillance, the U.S. Navy is well positioned to expand its efforts in this field beyond the protection of fleet assets to include the broader information collection and sharing goals of the MSP initiative.

Again, the promise of integrating such capabilities into a broader maritime surveillance regime appears to warrant further investigation.

Enhancements of Existing and Planned National Technical Means/National Reconnaissance Office (NRO) Satellites. In recent years the main threats against which the United States has postured its military capabilities in general and its orbiting surveillance capabilities in particular have been terrestrial. Consequently there has been little apparent attempt to extend the capabilities of these enormously capable and expensive NRO systems to improved surveillance of important ocean areas in ways that would help locate and identify uncooperative ships. The committee believes that the prospect of large payoffs for relatively small enhancements of the functionality of planned new spacecraft are sufficiently attractive to warrant a detailed investigation of the cost and probable effectiveness of such enhancements.

Findings and Recommendations

The U.S. Navy is uniquely qualified to help expand international maritime surveillance in support of its and its partners' maritime security goals. In particular, as the nation's primary repository of expertise on broad ocean surveillance, the U.S. Navy is best qualified to help improve the surveillance of key areas of the ocean surface and to provide the additional surveillance information to other nations whose maritime security it would enhance.

Finding: There is a range of technical options for improved ocean surveillance, some of them near term, that should be less costly than fielding large, new sensor systems. Some of them exploit data from a growing inventory of commercial remote-imaging sensors and satellites, others entail maritime-directed upgrades to existing over-the-horizon radars and/or national reconnaissance systems, and, finally, still others involve coastal radar surveillance of the near-in waters of partner states.

Finding: In many parts of the world, U.S. naval component commanders are well positioned to encourage coastal nations to improve their own maritime surveillance capabilities. To this end there are some relatively low-cost, high-payoff improvements for which the Navy could provide not only technical assistance (an example would be the selection and siting of coastal radars) but also material assistance by such means as the Section 1206 funding mechanism.[15] In some places such programs are well under way, but many more opportunities could be productively pursued.

Recommendation 8: The Chief of Naval Operations (CNO) should direct the Director of Naval Intelligence (N2) and the Deputy Chief of Naval Operations for Communication Networks (N6), and the Assistant Secretary of the Navy for Research, Development, and Acquisition should direct the appropriate laborato-

---

[15]Section 1206 funding, named for the section of the 2006 National Defense Authorization Act that authorizes it, is designed to help other countries build capacity within their national military forces. The authority allows DOD, in consultation with the State Department, to spend up to $200 million a year to help other countries.

ries, system commands, and program executive offices to increase their efforts to investigate, analyze, and help field, if appropriate, the most cost-effective combinations of capability across the potentially promising approaches to persistent, improved broad ocean surveillance that are identified in Chapter 3. To facilitate this initiative, the CNO should (1) seek a higher level of representation at the National Reconnaissance Office, where decisions are made on U.S. sensor performance goals, and (2) leverage its newly expanded role in the Office of the Director of National Intelligence (ODNI) to encourage the inclusion of maritime surveillance features in the next generation of commercial remote sensors from which the ODNI expects the agencies, particularly the nongovernmental agencies, to contract for products.

Recommendation 9: The Chief of Naval Operations, in coordination with the combatant commanders, should direct the Director of Naval Intelligence (N2) and the Deputy Chief of Naval Operations for Communication Networks (N6), and the Assistant Secretary of the Navy for Research, Development, and Acquisition should direct the appropriate laboratories, system commands, and program executive offices to ensure that naval component commanders have the appropriate expertise and other assets to facilitate an outreach program to coastal states that would benefit from improved maritime surveillance capabilities.

Analyze/Fuse

Framing the Challenges

Given the variety of current and potential surveillance data and intelligence information that can be exploited, it is important to recognize the functions that will need to be performed in the analysis and fusion of this multimodal data. As depicted in Figure 3.17, there are four such functions required to transform data into actionable decision-making information:

- Data conditioning is the development of a common ontology/data model.
- Data fusion entails combining data at various levels.
- Data mining involves the discovery of patterns and associations in large static datasets.
- Human–systems collaboration environments entail the development of visualization and collaborative decision-making tools for maritime security analysis and tasking.

For reference purposes, the Joint Directors of Laboratories (JDL) data fusion group model developed by the U.S. DOD JDL/Data Fusion Subpanel is cited. The five levels of fusion are (1) subobject data association and estimation: pixel/signal-level data association and characterization at the sensor level (L0), (2) object refinement (L1), (3) situation refinement (L2), (4) significance estimation or threat refinement (L3), and (5) process refinement: adaptive search and processing resource management (L4).

For the foreseeable future, the technical challenges of data mining and data fusion will be heavily concentrated on the U.S. side. Clearly, the multinational shared information space—lower volumes from fewer sources of raw data, at least to start—will not demand this level of information analysis and exchange. However, overall data mining, analysis, and fusion technologies must advance in order to create rationalized alerts and actionable information that may then be shared with partner nations in sanitized form in accordance with bilateral agreements. As regional communities share information with the global community, the amount of multimodal data that will need to be mined, exploited, and shared in a timely manner in order to maintain maritime security will pose several technology challenges in each of the MDA analysis/fusion functional areas.

Data Conditioning Central to the analysis and fusion problem is the need for a common language to describe the data in the context of maritime security. Data from disparate sources must be translated into a form that can be cross-associated, time-stamped, and/or correlated by the fusion and mining components. This data alignment challenge grows as the number and variety of data sources increase. And, as the data set expands to include internationally gathered information, special consideration must be given to language translation technologies such as those used in Translingual Instant Messaging (TRIM), employed in the CNIES program by JIATF-S, or in the Foreign Language Media Monitoring program sponsored by the Defense Advanced Research Projects Agency (DARPA).

Overall, the committee notes that some good technology work is being done in this area; however, most of the efforts are focused on a specific regional MDA solution or a unique data conversion challenge. There is no broadly applicable maritime domain data model with associated translation technologies. However, the efforts of the MDA DS COI are an important first step in the development of an overall architecture for information management and dissemination in this domain. More specifically, the MDA DS COI effort is working to transform data discovery and access from stovepiped systems to Web services using DOD Net-Centric Enterprise Services and the MDA COI standards. The envisioned capability is aimed at full MDA data exposure at unclassified and GENSER levels based on the MDA COI schema.

Data Fusion As shown in the diagram, data fusion as defined here involves (1) sensor data fusion, (2) anomaly detection, and (3) vessel context association (incorporating data such as crew, cargo, financial data, ownership, and so on).

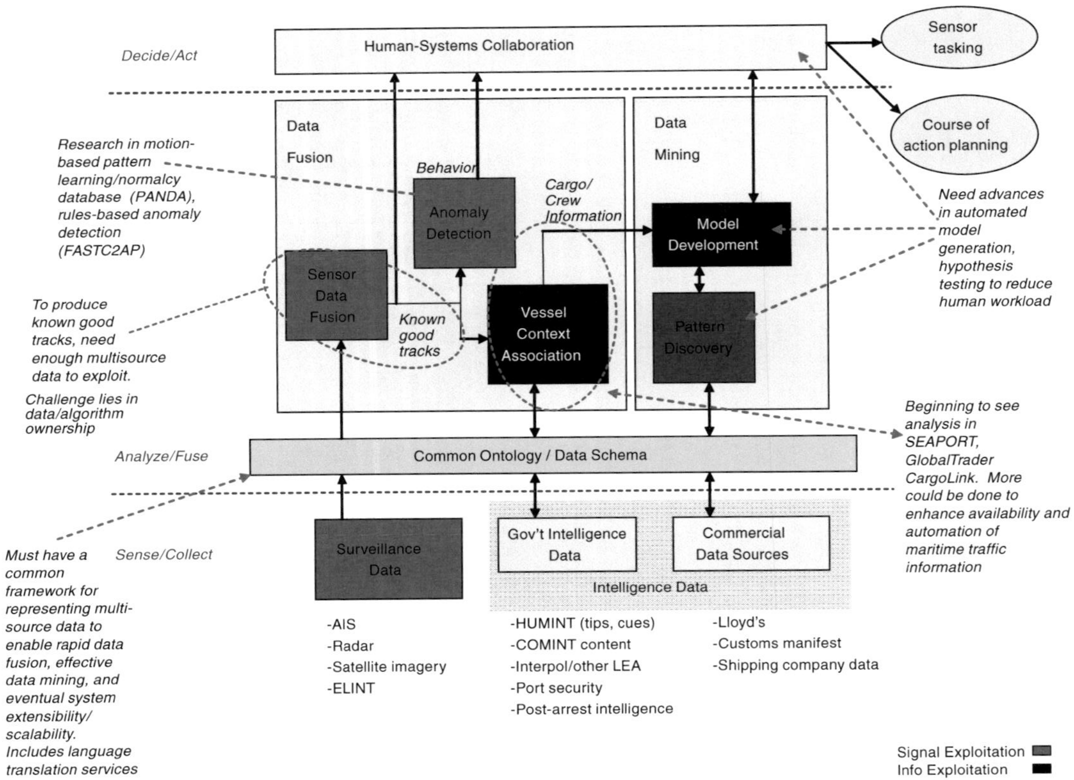

FIGURE 3.17 A functional view of maritime domain information analysis/fusion.

Sensor Data Fusion. This capability involves correlating available sensor information (emitter characteristics, imagery, AIS data, and so on) to positively identify a maritime entity and associate that entity with track data. To produce known good tracks (i.e., accurate positions and unambiguous updates) and subsequently detect anomalies (i.e., vessel behavior outside the norm), there must be enough multisource data to exploit. With the development of more data types and data sources to contribute to MDA (as described in the section "Sense/Collect"), and if the above recommendations for analysis and placement of surveillance sources with adequate sensitivity and coverage are exercised and assets are directed and utilized appropriately, the availability of data should not be an issue in the future. Instead, the challenge will lie in how well all the available information can be fused together into high-quality, multisource tracks, preferably including vessel, crew, and cargo information. Rising to this challenge will require resolving competing claims to ownership of the data and the algorithms that process the data.

Anomaly Detection. The U.S. maritime security community currently monitors dozens of high-interest vessels at any one time that could be affiliated with entities or individuals that could be involved in terrorism or other threatening or illicit activity. As noted in an earlier section, the identification of vessels of interest is substantially dependent on tips, often from law enforcement sources. Information on cargo and crew, for instance, is analyzed as part of the process. Such analysis can be quite effective and will continue to be important. However, this analysis can involve a substantial amount of cognitive effort by those who stand watch and by analysts to infer intention or activity from patterns of motion and other observables. The purpose of the various anomaly detection initiatives is essentially to automate this manual process and enable early detection of an emerging threat. The current focus is on identifying threat vessels from among the more than 50,000 vessels over 300 GT that are engaged in international maritime commerce. Such an automated capability could then bring vessels of interest to an operator's attention for manual verification and a decision on course of action. An essential step for the development of a robust, automated process is the understanding of common practices of the maritime community and the representation of that understanding in a normalcy database.

DARPA research in programs such as the Fast Connectivity for Coalition Agents Program (FASTC2AP) and Predictive Analysis for Naval Deployment Activities (PANDA) is advancing the technology used to predict potential threats. For example, FASTC2AP is designed to allow users and watch-standers to specify vessel behaviors and characteristics that drive alerts and prompt operators to analyze those vessels further. Essentially a human-interactive, rule-based program, FASTC2AP represents a first step toward automated anomaly detection. For example, an analyst can specify a set of known suspicious behavior patterns (deviation from sea lane, slow speed in sea lane, close proximity/loitering with another ship, and so on) and specify rules that will trigger alerts.

The PANDA program is aimed at developing a normalcy database for a given ship's motion and using that normal behavior model to automatically identify anomalies or potential threats. Detection of potential threats based on analysis of vessel motion augmented with emission analysis is within the realm of possibility due to the increasing amount of information being collected by port authorities (by using human intelligence and law enforcement agencies), by the AIS, by other open sources, and by intelligence community sources that provide persistent, long-duration tracks on many surface vessels. The problem is highly challenging not only because there are so many monitored vessels but also because multiple organizations—continental United States (CONUS)-based intelligence centers, fleet-level fusion centers, and shipboard situation awareness cells—have separate roles, sensor ownership, and data access. Hence, any solution needs (1) to enable decentralized analysis; (2) to scale to different data availabilities, data rates, and levels of resolution; and (3) to permit efficient exchange of models, tracks, predictions, and alerts across organizations and classification levels. Essentially a fusion program, PANDA focuses on predicting vessel behavior given known good multisource track data, identifying potential threats, and cueing the further exploitation of the information (crew, cargo, financials, and so on) that is correlated with that vessel.

Both FASTC2AP and PANDA are research programs that will require a transition sponsor should they prove to be effective. FASTC2AP is at a technology readiness level (TRL) of 9 and is installed and has been used operationally by the Sixth Fleet and by NATO maritime elements. The goal is to harden the technologies and transition them to Navy programs of record by FY08-FY09. PANDA is in its first year (of four) as a DARPA program. A transition sponsor is to be determined.

Moreover, the products of these programs, while valuable to U.S. interests, will probably not be available to all the regional partners with which information is often shared at the unclassified level. On the other hand, the committee's view is that selected basic capabilities can and should be provided to partner nations to advance both trust and capability. A basic analysis/fusion tool set such as might be included in a starter kit could allow the examination of more anomalies within such regional networks. A first anomaly detection capability might be software that can detect a change in reported AIS identification for a given vessel or that can detect a vessel traveling in a manner not consistent with its destination. In addition, and apparently as was envisioned as part of the noted MDA spiral 1 effort, selected vessel tracking algorithms can be shared as releasable GOTS.

Although they are well worth pursuing, anomaly detection schemes are likely to have only limited ability to identify vessels of interest among the large numbers of smaller ships whose historical behavior has been noisy—that is, ships that have no stable historical pattern against which current behaviors can be compared.

Beyond Vessel Information. Clearly, much of maritime security involves tracking of people and cargoes. As evidenced in JIATF-S operations, that class of information is absolutely crucial to success and requires an extension of interagency relationships to better track the flow of people and cargoes of interest and relate them to shipping activity. Similarly, efforts in both the law enforcement and intelligence communities involve linking vessel track data with the information on a ship's manifest to identify vessels of interest. To reduce the threat to maritime security, emphasis must be placed on analyzing and selectively sharing information on crew, cargo, supply chain, financials, ownership, and so on. The committee understands that access to this information implies the existence of a robust interface with the broad law enforcement community such that relevant vessel context information can be fuzed with vessel track and behavior predictions. Linking this context information is automated by the use of tools such as those sponsored by the Office of Naval Intelligence (ONI) and investigated as part of the N6 MDA activity. For example, one such tool, SeaPort, is establishing a global MIO database. Another tool, Global Trader, is currently focusing on foreign-to-foreign transport of shipping containers (60 to 70 percent of all global shipping data). Another unclassified module, Cargo Link, is expected to provide for research and prediction of cargo data and pattern mapping for cargo and manifests.

Data Mining Data mining involves correlating various kinds of information from both structured and unstructured databases and evaluating the correlated data sets to detect previously unknown patterns. It focuses on the relationships between objects in the databases and involves link analysis, which is building networks of interconnected objects in order to predict future events. The main tasks of link analysis are to extract, discover, and link together sparse evidence from vast amounts of data, to represent and evaluate the significance of the related evidence, and to learn patterns to guide the extraction, discovery, and linkage of entities. The discovered relationships may be transactional, geographical, social, or temporal.

Data mining technologies of this sort have been applied in a number of domains—in commerce, stock-market analysis, medical research, and the insurance industry, to name a few. As a result, there are countless commercial tools available to help analysts in a variety of fields absorb a vast amount of structured and unstructured data, visualize patterns, and predict behavior based on the patterns that the analyst uncovers.

While there are many commercial and custom tools to perform this pattern recognition, few tools exist to perform pattern discovery, specifically in the area of maritime security and threat analysis. Advances in automated model generation and hypothesis testing are required to reduce the human workload and expedite mining through the additional information that is likely to result from data sharing with nations in the MSP. The approaches under investigation appear consistent with the Internet search architecture, although perhaps only a subset of such mining tools may be releasable to regional partners for unclassified databases.

Human–Systems Collaboration Environment Visualization and analysis tools that allow an analyst to connect the dots (derive vessel intent, postulate threat scenarios, and so on) are an essential element in creating an MDA picture. Given known good tracks, predicted vessel activity, and maritime behavior models, command and control centers should be able to understand where—that is, on which vessels of interest—to focus their maritime security operations. However, there is a great need here as well for advances in automated model generation and hypothesis testing to further reduce the human workload. Current data mining and visualization techniques will have difficulty keeping up with increasing amounts of MDA-related structured and unstructured information.

The state of the art in MDA data fusion and data mining requires much interaction with humans. Since most technology solutions in use today solve the level 0 (data source processing) or level 1 (object refinement) problem, no real capability has been implemented to handle fusion for the large amounts of diverse, uncertain maritime data at higher fusion levels. Indeed, automated fusion tools at JDL levels 2 (situation refinement) and 3 (threat refinement) are part of the science and technology (S&T) community research agenda. DARPA's FASTC2AP program, for example, includes development of Web-enabled tools for global maritime awareness. FASTC2AP's human–machine interface allows users to create and configure agents and deploy Web portal technology to automate the currently manual processes. While there is some specific research work in this area, efficient and effective integration of humans into the fusion process is not yet widely understood.

Higher-level decision support functions (detecting anomalies, predicting behavior, establishing relationships, deriving intent) is, as a result, primarily performed by analysts. To make good use of increasing amounts of diverse maritime-related data, research (and plans for subsequent technology transition) in human-guided fusion algorithms and automated fusion technologies must be included in an overall MSP strategy. More specifically, research is needed to take maritime situation awareness to the next level of data visualization, relationship exploration, and link analysis in order to strengthen maritime intelligence discovery.

Advancing the State of the Practice

Today, the majority of maritime domain information is in stovepiped systems, and the focus is on tracking vessels of interest. While this manpower-intensive practice is useful in reducing the potential for harm by terrorists or criminals, more can be done through increased data collection, analysis, and fusion to further reduce this threat. As illustrated in Figure 3.18, future MDA capability could further support common maritime security interests, provided that (1) there is a common baseline for situation awareness at the unclassified level and (2) current research and demonstration technology aimed at higher levels of fusion is successfully transitioned into the operational community. Although the figure focuses on the terrorism threat, it depicts principles that apply more broadly.

Providing Shared Analysis and Fusion Capabilities in the Near Term One way to advance the state of the practice is to start with a baseline for a well-understood maritime COP. In other words, we can advance the state of the practice by bringing all partner nations up to a base level of situation awareness through the dissemination of starter kits that include COTS or releasable GOTS fusion tools. These tools for merging AIS, imaging systems, and radar data would create an integrated and more reliable situational awareness capability for all MSP participants. Candidate starter kit tools can be provided that have been quite well

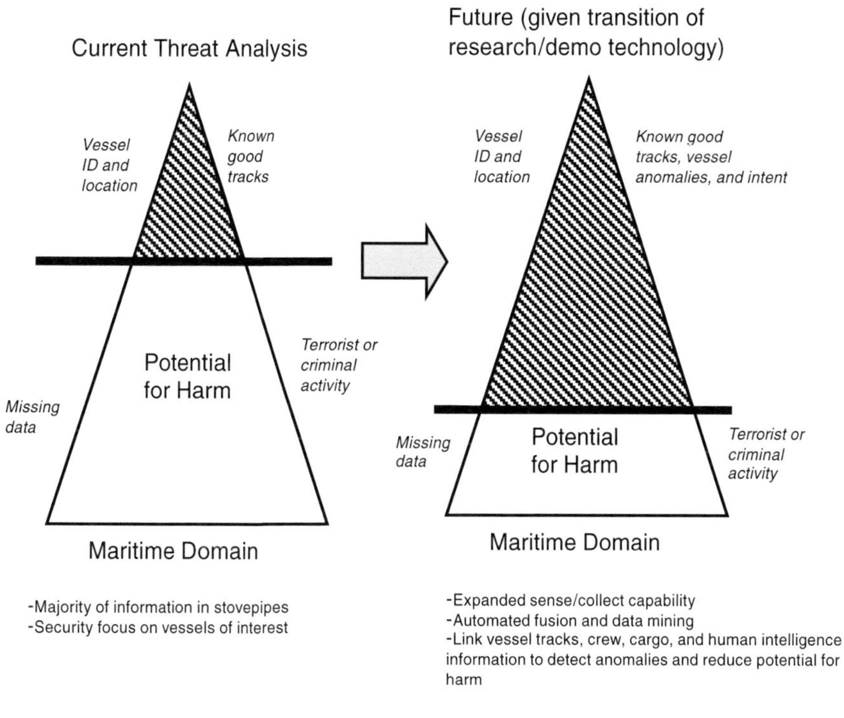

FIGURE 3.18 Strengthening maritime domain threat recognition. SOURCE: Based on Chris Dwyer, Naval Research Laboratory, 2007, "Comprehensive Maritime Awareness (CMA) Joint Capabilities Technology Demonstration (JCTD)," *Proceedings of SPIE* [Society of Photo-Optical Instrumentation Engineers], Vol. 6578, Defense Transformation and Net-Centric Systems 2007 [Conference], Orlando, Fla., April 9-13.

tested in operational implementations; they would be used primarily for vessel identification and tracking. Consider, for example, NavAir's SureTrak, a proven MDA system capable of interfacing with existing sensor and C2 systems. Using data from a variety of sources (surface and air surveillance radar, video tracking systems, GPS, AIS, and so on), SureTrak is used primarily to improve harbor surveillance. This computer display system monitors marine harbor traffic, issues advisories to vessels in areas elected by the system operators, and provides the operators of the system with an early warning of unacceptable traffic conflicts in the confined waterways of the harbor. Each system consists of a number of remote sites providing radar, camera video, and audio communicated to a central vessel traffic center (VTC). VTC data integration and display provide the ability to identify and monitor vessel traffic by fusing multiple radars on a single display. SureTrak is a sample open-architecture, government-owned, and commercially available fusion system. The U.S. Navy should consider sharing some of these technologies at the unclassified level with nontraditional partners in order to boost rudimentary MDA. SureTrak and other such technologies are already being considered as part of the Navy MDA activity.

Commercial tools for data fusion and data mining might also be of interest and should be considered a component that requires the participation of humans. Analyst toolkits designed to perform statistical processing would be of little value to a Navy operator; however, tools to associate data from disparate sources, threat and risk assessment tools, and collaboration and visualization tools could certainly augment the overall fusion and analysis capability.

Leveraging Advanced Technology Research The current focus appears to be on an evaluation of the available technology (both COTS and emerging GOTS) in order to solve first-order MDA problems (vessel tracking, rudimentary anomaly detection, threat identification, communications, and collaboration). This approach strikes the committee as the right course of action to begin development of an integrated MDA capability. However, significant additional system design and engineering will be required to develop an intelligent, integrated, and automated MDA COP.

At JDL levels 0 and 1, technologies exist and can be integrated into a prototype solution insofar as they have been developed for incorporation into a net-centric, service-oriented architecture. On the other hand, technology at higher JDL levels is currently a more advanced research problem, as discussed above. Higher levels of fusion technology are being developed under various S&T programs sponsored by research-oriented government agencies. The aforementioned DARPA programs (PANDA, FASTC2AP) are examples of the results of such fusion research. In addition, the Office of Naval Research (ONR) has taken the initiative to leverage commercial technology in such programs as Pattern Analysis and Bayesian Link Discovery Tool for Transactional Networks (PALADIN) and Cleverset. Through ONR's Commercial Technology Transition Office, PALADIN was developed to detect threat activity and perform network analysis by efficiently searching massive, noisy data that may be unreliable, incomplete, or inconsistent. PALADIN's anomaly detector, partial pattern matcher, hypothesis evaluator, and hypothesis merger can be used for maritime data fusion and analysis. They include a data model and database interface specification for extracting entities, links, and attributes from new data sources. PALADIN also includes a network visualization tool for exploring and discovering networks and evaluating threat hypotheses. In another ONR-sponsored example, Cleverset was given a small business innovation research award to apply its commercial algorithms in the development of improved report-to-track (RTT) fusion, track-to-track (TTT) fusion, and hybrid RTT/TTT fusion technology. Through both government-sponsored research and commercial technology transition, the gaps in MDA can begin to be filled.

In addition to needing funding and transition agency sponsorship, technology transition cannot occur efficiently without software engineering for the research work products. The DOD vision for net-centric warfare will necessitate a fresh view of the software engineering—that is, of the packaging of these emerging fusion technologies. For these technologies to fit into a net-centric information exchange environment, a common lexicon, software framework, and protocols must be developed and leveraged. In some cases, legacy technologies developed for closed-environment, stovepiped systems will need to be reengineered to ensure that the newly refactored software encapsulating a custom fusion algorithm is extensible, modular, portable, and self-describing. Clearly, a strong emphasis on software architecture and meta software project management must be considered in any large-scale systems integration effort for MSP. This is evidenced in the N6 MDA prototyping efforts, where the criteria for incorporating a technology into the demonstration spirals include not only the technology's ability to address the requirements for MDA but also the ease with which the technology fits into a net-centric, service-oriented architecture. In short, if the technology is not or cannot be packaged correctly to fit, transition will be severely inhibited.

Findings and Recommendation

Finding: Research and demonstration programs sponsored by various agencies have produced good work that addresses some of the technology gaps in the current analysis and fusion of maritime domain awareness information. Much of the technology being developed to analyze and fuse data on maritime entities is in the early stage, in prototype form. However, as reflected in Navy efforts ongoing as of this writing, there are commercial off-the-shelf and potentially releasable government off-the-shelf analysis and fusion tools and software that offer early, useful capabilities for maritime security partnerships.

Many ongoing maritime security and domain awareness efforts are currently funded under Iraq and Afghanistan supplemental budgets. This situation could result in an unfortunate loss of focus on MDA and a loss of momentum in the development of an overall MDA architecture when supplemental budgets for these nonmaritime contingencies wind down or stop.

Recommendation 10: To leverage analysis and fusion technology and tools, the Chief of Naval Operations should assign the Deputy Chief of Naval Operations for Communication Networks (N6) (along with the relevant laboratories and systems commands) to take responsibility for maritime domain awareness-related analyze-and-fuse technologies, either for their short-term application as part of a starter kit (in releasable government or commercial off-the-shelf form) or for longer-term advanced research with identification of transition opportunities. Given that these efforts are of long-term importance, independent of the purposes of current supplementals, the Deputy Chief of Naval Operations for Resources, Requirements, and Assessments (N8) should work on funding maritime domain awareness efforts in the mainstream of the Navy budget.

Recommendation 10 expands Recommendation 7, which calls for the development of IT infrastructure starter kits to facilitate and accelerate operational information-sharing initiatives that include analysis/fusion tools.

Decide/Act

As described above, the decide/act function calls for consultation and coordination mechanisms to support the decision process among partners and the execution of an action once a decision has been made. There is a very broad range of appropriate responses to the detection of suspicious activity and the sharing of the information. Such responses range from a simple maritime intercept operation by the patrol craft of a coastal nation in response to information provided by another nation, to highly complex, coordinated multinational use of aircraft, ships, and port authorities to deal with a suspected perpetrator.

Realizing the ultimate benefits of new or strengthened partnerships demands that such mechanisms exist at both the operational level (among partner nodes and centers) and the tactical level (among ships, boats, and aircraft and their command nodes). Key enablers include the following:

- Bilateral or multilateral agreements that specify the allowable scope of action, the rules of engagement, and so on. Table 3.2 illustrates some tactical actions that are codified in existing bilateral agreements, in this case a template of eight possbile actions taken from agreements between the USCG and partner nations in the south Atlantic and Caribbean (Coast Guard District 7);
- Supporting procedures—for example, a partnership analogous to the U.S. interagency Maritime Operational Threat Response conference procedures developed in response to the NSMS); and

• Supporting system capabilities, which are the focus of this section, "Decide/Act."

Several building blocks of the supporting system's technical capabilities can be identified. For coordination and consultation, there is a substantial amount of readily available COTS tools/functionality—and sometimes even releasable GOTS—in areas such as multimedia collaboration and multilingual chat. The TRIM tool used by JIATF-S for Spanish translation—but supporting some 13 lan-

TABLE 3.2 Representative Tactical Action Agreements

Tactical Action	Tactical Action Agreement
Ship boarding	Standing authority or procedures for the USCG to stop, board, and search foreign vessels suspected of illicit traffic located seaward of the territorial sea of any nation.
Ship riding	Standing authority to embark law enforcement (LE) officials on platforms of the parties, whom officials may then authorize to perform certain law enforcement actions.
Pursuit	Standing authority or procedures for U.S. government LE assets to pursue fleeing vessels or aircraft suspected of illicit traffic into foreign waters or airspace. May also include authority to stop, board, and search pursued vessels.
Entry to investigate	Standing authority or procedures for U.S. government LE assets to enter foreign waters or airspace to investigate vessels or aircraft located therein suspected of illicit traffic. May also include authority to stop, board, and search such vessels.
Overflight	Standing authority or procedures for U.S. government LE assets to fly in foreign airspace in support of counterdrug operations.
Relay order to land	Standing authority or procedures for U.S. government LE assets to relay an order to land in the host nation to aircraft suspected of illicit traffic.
International maritime interdiction support	Standing authority or procedures for U.S. government LE assets to moor or stay at national ports, entry of additional U.S. government LE officials (by ship and/or aircraft), entry of suspect vessels not flying U.S. or host nation flag, escort of persons from suspect vessels through and out of host nation (by ship and/or aircraft), and landing and temporarily remaining at international airports for logistics.
Third-party platforms	Provides for operations from vessels of nations other than the parties to the bilateral, usually by LE detachment from third-party vessel.

guages—is an example of releasable GOTS. Such software offers support at both the operational and tactical levels and calls for only a modest PC capability.

For the exercise of C2, there is the obvious need for connectivity extending to the tactical level. The solutions range from a rudimentary, beyond-the-line-of-sight radio voice capability to high-bandwidth, satellite-based data capability. Here, too, capability building blocks are readily available. For instance, Navy plans are leveraging CENTRIXS-provided capabilities and call for providing Iridium satellite phones to selected partner nodes as part of a fly-away package (see Figure 3.6). As mentioned earlier, DOD makes available a GOTS PC-based C2 package suitable for supporting these types of activities.

Technology opportunities exist in this functional domain, too. For example, beyond the current technology for video teleconferencing, an emerging so-called telepresence technology is beginning to provide realistic and full contextual face-to-face experience. Further, the section "Analyze/Fuse" touched on the technologies and decision-support tools in areas such as visualization.

Clearly, the selection of technologies and fielded products must be tailored to the supporting infrastructure, defined broadly—for example, bandwidth (the well-known "disadvantaged user" issue) and sustainment and training capabilities.

Providing collaboration, consultation, and coordination capabilities at the operational and tactical levels is not viewed as a complex technological challenge. The issues involved in sharing such support with nontraditional partners relate to the availability of COTS or releasable GOTS products and tailoring them to the situation at hand. The provision of communications and collaboration tools and systems should be included within the broader "design template," "maritime security partnerships catalog," and "starter package" referred to in Recommendation 7.

PROTECTING WHILE SHARING INFORMATION

The concept of MSP requires the collection, storage, and sharing of information, but the potential for disruption and compromise exists at each of these stages. Depending on a number of factors, including level of trust, potential vulnerabilities, and cost and availability of information protection solutions, different connectivity architectures will be employed for different partnerships. In addition to the concerns inherent in maintaining secure communications and networks, there is the issue of protecting the information itself, with concerns ranging from revealing sensitive ship positions to giving away a competitive advantage. These are concerns for both the United States and the prospective maritime partners.

The approach to assessing potential vulnerabilities when sharing information starts with an (open source) assessment for different levels of connectivity among the partners and is followed by a generic assessment of the vulnerability of the systems architectures envisioned to support these partnerships. This is followed by an assessment of residual vulnerabilities and their impact on the sometimes difficult trade-offs between sharing and protecting information within a partnership context.

Figure 4.2 in Chapter 4 lays out a spectrum of maritime security issues, from traditional military naval warfare at the high end to law enforcement issues such as illegal fishing at the low end of conflict. The issues will be resolved by different information security and protection regimes found across this spectrum. For example, U.S. ties to its closest allies deal with the entire security spectrum and often involve the sharing of Secret information (e.g., CENTRIXS networks), while its less mature partnership arrangement might involve sharing unclassified information, perhaps including sensitive law enforcement information, at the lower end of the spectrum.

General Considerations

Box 3.1 pairs the sources of potential threats to MSP and the tools they use to exploit system vulnerabilities. Although MSP does not think of nations per se as the only potential adversaries, a nation might be suspected of engaging in a hostile act if it were perceived to be behaving counter to its own interests in matters such as fishing rights, navigational freedom, or environmental restrictions.

Certain competitor nations have highly sophisticated capabilities in information operations, but the risk that they would mount an all-out attack on MSP information systems appears to be slight. On the other hand, a national power might wish to obtain or compromise MSP data to gain a commercial advantage.

Terrorist and criminal organizations can hack into computer systems to steal information, alter databases, and disrupt networks. It is assumed that they would use these capabilities sparingly since their principal objective is to avoid detection. The main concern is their acquisition of privileged information.

The potential exists for nonstate actors to disrupt partnerships for political or ideological purposes. Such hackers or activists have demonstrated the ability to disrupt major networks with distributed denial-of-service attacks.

BOX 3.1
Hierarchy of Threats and Vulnerabilities to MSP Connectivity and Information Protection

- National (e.g., North Korea)—information operations, physical attack
- Terrorist organization (e.g., WMD transport)—hacking, deception
- Criminal organizations (e.g., drug cartels, piracy)—hacking, deception
- Nonstate actors (e.g., hackers and activists)—network attack
- Legal "infringers" (e.g., fishing rights, immigration)—deception

At the low end of the threat spectrum are violations of a partner's laws or rights, such as happens with illegal immigration or an encroachment on fishing rights. Many countries seek partnerships with the United States because it is in their interest to do so. The threat perpetrators pose to connectivity and information protection is minimal because, again, their priority is to remain undetected.

The threat from an MSP standpoint is the potential for compromise of information that partner nations wish to keep private from nonpartner entities for reasons of national security or commercial advantage. If partners feel their information is not secure from unauthorized access or intentional data corruption, they may decline to share it. While breakdown in connectivity is a possibility that cannot be overlooked, it appears to be less of a threat to privacy.

Protection Technology

Table 3.1 listed seven systems that enable maritime information sharing, and Figure 3.7 depicted the N6 multilevel sharing architecture from unclassified systems such as MSSIS to classified systems such as CENTRIXS. Information protection issues exist with the sharing of unclassified as well as classified information (e.g., CENTRIXS nets for Joint Task Force-150). For instance, law enforcement information related to tips is generally viewed as sensitive even though the information has not been classified in a formal sense. Given the range of security regimes driven by sharing at different levels of classification and/or sensitivity, it is important to identify a corresponding range of readily available building blocks for information protection.

The architecture for information sharing between or among nontraditional partners will be implemented with COTS products integrated into an open architecture backbone context and protected by COTS security products. Classes of information and network protection technology are listed in Box 3.2.

BOX 3.2
Classes of Information and Network Protection Technology

- Multiple security levels (not the same as multilevel security)
- Commercial security technology
  —IP Sec (IPv4, IPv6)
  —Secure Sockets Layer, Virtual Private Network
- Multilevel security technology
  —Hardware-enforced security
  —Software-enforced security
  —Radiant Mercury
  —Trusted operating systems
  —Guards

Multiple security levels are required for protection of classified information as opposed to software-imposed multilevel security in an operating system. As an alternative to human-intensive "air gaps" to protect information and networks on the U.S. side of the interface, automated, filtered interfaces (e.g., Radiant Mercury guard) between security levels are needed to ensure capacity and timely workflow. Issues exist with current guard technology and products. For example, Virtual Private Network (VPN) security via commercial Internet service provider connection is blocked by some routers if Network Address Translation is applied behind a firewall to increase the number of users at a single IP address. However, these issues can be overcome with proper system design.

U.S. policy with respect to protection is driven by the level of protection associated with the information. Some national security information can be deemed to be classified and possibly also compartmented. Other national security information can be deemed to be unclassified, with a wide range of Controlled Unclassified Information (CUI) designations, including relevant law enforcement information standards. Finally, some unclassified information is not considered as national security information but yet may require protection under a particular partnership agreement.

Decisions made with respect to a particular partnership arrangement within which various kinds and levels of information are to be shared will dictate policy and derivative requirements for certification; acceptable choices among protection strategies; and products in areas such as user authentication, access controls, and information confidentiality.

In the maritime sharing domain, concerns may arise about aggregate ship position information, which might compromise competitiveness, or about the potential exposure of law enforcement sources and methods. Therefore, even for unclassified information, commercial security such as Type 3 encryption, VPN, SSL, and Transport Layer Security (TLS) would be appropriate. Other commercial products for security and control of access to information include ID cards with biometrics for user authentication.

Even networks and databases handling unclassified information need consistent application of COTS privacy and security products. DOD policy, although apparently not uniformly enforced, is that so-called common-criteria products certified by the National Institute of Standards and Technology (NIST; not the National Security Agency) are used in such cases. A difficulty with the NIST-certified products is that it costs vendors time and money to get certified, so the number of available building blocks is constrained.

Managing Risks

The application of an open architecture employing commercially available security technology basically ensures that there will be some degree of vulnerability for system and data integrity. In general, then, the issue here is one of

BOX 3.3
Information and Network Security Vulnerabilities

- Insider threats
- Directed denial-of-service attacks
- Hacking (malicious code, interception of data, insertion of false data)
- Jamming
- System breakdown
- Lack of configuration control (loss of interconnectivity)
- Unintended recipients of information

managing risk. It must be assumed that some of the shared maritime information will somehow become available to adversaries of the United States and its partners, including terrorists and criminal elements, through insider knowledge if not through network penetrations.[16]

Box 3.3, a listing of residual vulnerabilities assuming the application of commercial security protection, includes vulnerabilities associated with adversarial actions but also includes system design-level vulnerabilities that can bring down networks and compromise information. The most common forms of computer network attack are to overload the network to bring it down (distributed denial of service [DDOS] attacks) or to somehow gain access to the system (by hacking) to attack the operating system, create zombies, intercept data, or insert false data. Since ships require electromagnetic propagation for surveillance and connectivity, their transmitted signals are subject to interception and jamming. Commercial business practice is to release new code early and apply patches as bugs are found in the software. Hackers have become very adept at exploiting bugs before the patches are applied.

System-level vulnerabilities can also be anticipated if there is no configuration control. This issue can be addressed by U.S.-issued fly-away communication kits but would be a potential problem with partner-furnished equipment unless common standards for security products and their use are set. Unclassified information that provides information to low-end threats and assists them in avoiding detection may be broadcast. As a simple example, ship radars provide an early warning system for other ships equipped with simple radar detectors. However, for all other communications and data storage for unclassified networks, the committee foresees the common application of commercially available security products and practices.

[16]Peter A. Loscocco, Stephen D. Smalley, Patrick A. Muckelbauer, Ruth C. Taylor, S. Jeff Turner, and John F. Farrell. 1997. *The Inevitability of Failure: The Flawed Assumption of Security in Modern Computing Environments*, National Security Agency, Fort Meade, Md.

Summary

The committee recommends the use of commercial products and network principles for information protection when sharing with and among partner nations at the unclassified level. Recommendation 7 assumes the Internet poses vulnerabilities associated with security. Commercial technologies exist to handle lower-end protection and are being extensively used by the Navy. Automated means exist to transition information to different levels of security for association and fusion, but these are cumbersome and limited. In addition, there is technology that allows data to be stored, communicated, and processed by a multilevel security approach.

Despite the application of security technology, skilled opponents, design and configuration flaws, and equipment breakdowns will allow residual vulnerabilities. In particular, the insider threat is very difficult to prevent. The global business communities, such as banking, live in this environment and despite threats and occasional compromises continue to operate. Partnerships, particularly those dealing at the levels of sensitive but unclassified or controlled and unclassified should be able to operate in the face of an occasional compromise of information by criminals. Backup connectivity should be considered to maintain a sufficient level of trust with partners when the system is disrupted. The bottom line is that vulnerabilities will exist but are not seen as showstoppers for the overall concept of maritime partnerships. Risk can be managed by carefully selecting the information to be shared and adopting adequate protection measures.

The committee strongly endorses the Navy's adoption of commercial protection technologies and products, as evidenced in emerging partnership initiatives. However, in this area and the area of networking infrastructure, there is a need to identify and test solutions and to attend to the devil-in-the-details issues inevitably associated with their integration into a working system. The committee did not, however, find any signs of an end-to-end information protection analysis, nor did it observe a NIST certificate for any information systems.

Recommendation 7, which called for Deputy Chief of Naval Operations for Communication Networks (N6)-led architecting, engineering, and fielding service in support of operational initiatives, covers information protection technologies and products. In addition to developing an MSP catalog of tested products and related starter kits, technical efforts should include an end-to-end information protection analysis to ensure that the protection meets the expectations of the partners for the several networks in operation or under development.

STRENGTHENING AND ACCELERATING PARTNERSHIP OPERATIONS AND INITIATIVES—MISSION-DRIVEN SYSTEM ENGINEERING AND ANALYSIS

The Case for Broad-Based System Engineering and Operations Analysis

Beyond the technically based efforts to ultimately field the enablers discussed in the three functional areas discussed above, there are the system-of-systems or enterprise issues associated with (1) maximizing capability and performance of existing systems and assets, (2) identifying capability gaps and solutions for filling them, and (3) exploring the difficult trade-offs between capability choices in a constrained funding environment. For instance, the foregoing discussion of intelligence/surveillance identifies options for improving the maritime picture and the need to explore these, including, in the end, a prioritization of possible investments based on their contribution to operational mission outcomes. Further, there are choices to be made in all the functional areas. Is the return on a $1 investment in additional surveillance capability as high as the return on that same investment in better fusion and mining of information from existing sources?

The committee found fertile ground for mission-focused operational analysis during its visits and internal discussions. For instance, interactions with JIATF-S representatives clearly identified challenges associated with the allocation and deployment of scarce maritime surveillance and interdiction assets, a solid recognition and understanding of these issues on the part of experienced staff, and an intent to build a base of operationally oriented data for analysis (an "enterprise database"). However, operational imperatives understandably continue to dominate or even preclude substantive, sustained analytic effort. Figure 3.19 illustrates a case in point: an analysis of surveillance coverage performance for different combinations of assets over a representative search box, noting the broader question of allocating assets among the more than 3,000 such search boxes that make up the JIATF-S area of interest.

Even though such analytical challenges were not routinely discussed with presenters or during visits, they clearly exist wherever surveillance assets are being deployed and tactical actions are being taken and can be expected to persist as emerging partnerships mature. Furthermore, the pressure of day-to-day operational imperatives as partnerships mature is not viewed as unique to JIATF-S.

The idea is that providing operationally oriented analytical support to partnership operational elements in a responsive and tailored way would advance the cause of maritime security. The committee envisions that combining such analytical support with support for enterprise-level issues will result in a broadly based systems engineering and analysis activity in support of partnership operational elements. The systematic execution of such an activity calls for a mission-oriented framework of some kind that encompasses all of the functional elements in a mission; Figure 3.10 shows an example.

This mission-driven systems engineering and analysis would also accommo-

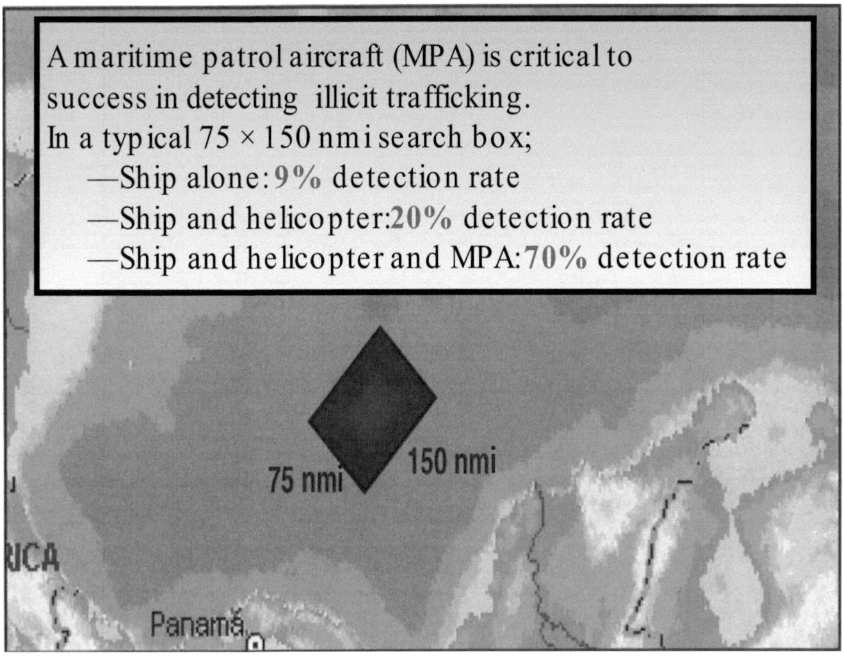

FIGURE 3.19 Allocation of surveillance assets to search boxes: a JIATF-S example. SOURCE: Joint Interagency Task Force-South, "The Importance of MPA," presentation to subgroup of the committee, June 12, 2007, Key West, Fla.

date planning for the future ("preplanned product improvements") and enabling technology developments and insertions. Examples include automated decision aids such as rudimentary anomaly detection.

A Corollary Effort—
Strengthening the International Maritime Security Regime

This report envisions the development of a two-pronged strategy for the building and strengthening of maritime partnerships—working regional and subregional initiatives and, at the same time, longer-term steps to strengthen international maritime security. Of particular interest here is the charter of the IMO, a central player in improving maritime security, and its successes in areas such as AIS and LRIT and in fostering standards for the reporting and exchange of relevant maritime information and working out agreements for the reporting and exchange procedures and obligations of its member nations. The committee believes that there are opportunities to extend and advance information reporting and sharing agreements that support maritime security and that the U.S. parties have opportunities to introduce constructive proposals and to support their further definition in an IMO working group. For example, one could conceive of reporting and sharing some classes of shipborne radar information, as discussed above, a topic that will presumably be addressed in an upcoming (as of this writing) IMO-hosted conference on such matters.

Technical analysis and support focused on topics like the relative merits of different data representation standards and mechanisms for collecting and sharing the reported information would of course be required. Such analysis and support is carried out today by the USCG as the U.S. representative to the IMO.

The extension of such efforts, as envisioned here by the committee, is motivated by the view that the United States could be more proactive in tabling proposals and driving them to realization, with technically based recommendations as a key element.

The Need for Technical Leadership by the Navy

Finding: There is a need—unsatisfied today—for a systematic, analytical approach to optimizing the design of the end-to-end system for the collection and analysis of maritime security information and its follow-up. Satisfying this need would require a range of technical support from the Department of Defense and interagency arena to foreign partners.

No matter how they are provided, support and advice should focus on system engineering for operational initiatives and would encompass related efforts such as the strengthening of U.S. technical participation in selected IMO initiatives as well as pragmatic, analysis-based advice to foreign partners on the most effective way to augment and deploy surveillance assets (e.g., radar siting).

Recommendation 11: The Chief of Naval Operations and the Secretary of the Navy should jointly propose a Navy-led and Navy-housed executive agent on the technical aspects of an information-sharing system for the U.S. interagency maritime security partnerships initiative. This agent would provide systems engineering and operations analysis resources with technical support to International Maritime Organization initiatives. This mission-driven, enterprise-level systems engineering and analysis capability would be an extension of the Maritime Domain Awareness Executive Agent role already assigned to the Navy by the Department of Defense. It would support not only the U.S. elements but also, under the auspices of ongoing initiatives, its foreign partners.

The enterprise-level systems engineering and analysis activity envisioned by the committee would address the following:

- Maximizing the capability and performance of existing systems and assets,

- Identifying capability gaps and solutions to bridge them,
- Exploring difficult cost/capability trade-offs,
- Allocating scarce assets to support operations,
- Mission-driven planning for future incremental improvements, and
- Identifying and planning for enabling technologies.

These activities would be accomplished from an end-to-end mission flow perspective, adopting an explicit framework for analysis (see Figure 3.9).

In this role, the Navy would be providing technical services to a range of customers: personnel at DOD, DHS, and at the Department of State elements responsible for leading and orchestrating MSP initiatives from a U.S. standpoint—for example, COCOMs, the USCG, and Department of State country teams as Navy's customers.

It is understood that the technical efforts envisioned here, to the limited extent that they are undertaken today, would be distributed among different elements across the Navy, DOD, and the federal agencies. However, the committee came around to the view that a serious commitment to the MSP concept calls for a dedicated system engineering and analysis activity postured to work on all the regional and subregional operations and initiatives. A dedicated, centralized activity would consider both user responsiveness and a mature center of excellence that serves as a repository for analytical tools used for the kinds of effort described here.

The committee understands that once such an effort is further defined and sized, it may well call for more funding than has so far been envisioned in MSP-related planning. At the same time as it realizes that new funding is always an issue, the committee also realizes that the funding requirements for the activity will probably be modest—a reasonable price for maximizing the mission performance of capabilities and assets involving substantially more investment and for informing decisions on future deployments and investments.

Looking Forward— An Interagency MDA Portfolio to Be Defined and Managed

The foregoing sections discussed system architectures and options for strengthening MDA information and its sharing in the 1,000-ship Navy context. Enabling management activities were called for in 11 recommendations. All of this, of course, implies investment. Just defining the options and assigning priorities is complicated by the fact that the MDA portfolio inherently cuts across multiple federal organizations and other systems (e.g., DOD, DHS, broader law enforcement, broader intelligence) and interfaces with international partner entities. The creation of the Director of GMSA position and the charter for GMII is of course designed to address the horizontal nature of the MDA challenge. The committee believes it would be highly desirable for the GMSA and GMII—with substantive support from the Navy as executive agent for the DOD—to take on the task of defining and establishing a management mechanism for the MDA portfolio.

Turning to the capabilities of interest and the Navy's investment therein, it seemed to the committee during its initial work that the Navy's focus was on exploiting the available information as much as possible (current dots) rather than, for instance, on seriously investigating potential new or enhanced surveillance capabilities, as outlined in this chapter (new dots). This focus and the resulting prioritization of modest resources seemed reasonable, and the reluctance to make potentially large investments in new surveillance systems without any clear and commensurate signs that they constituted a national security priority was understood.

Later on, as the committee was finishing its deliberations, the issuance of Navy guidance and the Navy's strong opposition to spiral 1 of MDA capability (the investment was apparently about $300 million) began to focus on and accelerate cross-community sharing and exploitation of information. Although the sharing was mainly with federal agencies as opposed to international partners, the committee viewed it as a very positive move.

Nonetheless, the committee remains concerned about the apparent lack of attention to strengthening maritime vessel surveillance. The idea here, reflected in recommendations in this chapter, is not that a large investment should be made in a particular system or capability but that a modest investment should be made now to explore in depth the full range of options, both those laid out here and others that will undoubtedly be identified. Known and potentially serious gaps exist in the technologies for active, assured surveillance. Clearly, promising options requiring significant investment would have to compete with other Navy and DOD needs.

In any event, the notion of a well-defined and actively managed MDA portfolio at both the interagency level and within the Navy is strongly endorsed by the committee.

4

Implementation Strategy for Maritime Security Partnerships

A clear understanding of the functions that must be performed to implement maritime security partnerships (MSP) and for which U.S. government executive departments or agencies have the responsibility and authority will facilitate that implementation. The functions are as follows:

- Creation of policy and strategic guidance that sets objectives, establishes regulations, and assigns roles, responsibilities, and authorities for conducting operations;
- Strategic and operational planning;
- Resource allocation;
- Development, management, and employment of the military and law enforcement forces; and
- Performance assessment and feedback.

THE NATIONAL STRATEGY FOR MARITIME SECURITY

The National Strategy for Maritime Security (NSMS)[1] provides broad strategic guidance for the development and coordination of MSP. It states as follows:

> The infrastructure and systems that span the maritime domain, owned largely by the private sector, have increasingly become both targets of and potential conveyances for dangerous and illicit activities. Moreover, much of what occurs in the maritime domain with respect to vessel movements, activities, cargoes,

[1]White House (George W. Bush). 2005. *The National Strategy for Maritime Security,* Washington, D.C., September.

> intentions, or ownership is often difficult to discern. The oceans are increasingly threatened by illegal exploitation of living marine resources and increased competition over nonliving marine resources.[2]
>
> Unlike traditional military scenarios in which adversaries and theaters of action are clearly defined, these nonmilitary, transnational threats often demand more than purely military undertakings to be defeated.[3]

It also calls for assisting partners to maintain their maritime sovereignty and jurisdiction over the seas. Along with safety at sea, the activities associated with NSMS reflect the activities for MSP. The NSMS calls for building partner capabilities. Moreover, it states as follows:

> Preventing unlawful or hostile exploitation of the maritime domain requires that nations collectively improve their capability to monitor activity throughout the domain, establish responsive decision-making architectures, enhance maritime interdiction capacity, develop effective policing protocols, and build intergovernmental cooperation. The United States, in cooperation with its allies, will lead an international effort to improve monitoring and enforcement capabilities through enhanced cooperation at the bilateral, regional, and global level, [by]:
>
> - Offering maritime and port security assistance, training, and consultation;
> - Coordinating and prioritizing maritime security assistance and liaison within regions;
> - Allocating economic assistance to developing nations for maritime security to enhance security and prosperity;
> - Promoting implementation of the Convention for the Suppression of Unlawful Acts against the Safety of Maritime Navigation and its amendments and other international agreements; and
> - Expanding the International Port Security and Maritime Liaison Officer Programs, and the number of agency attachés.[4]

The NSMS calls for new diplomatic initiatives through international organizations, coordinated by the Department of State, to include activities such as the following:

> - Implementing standardized international security and World Customs Organization frameworks for customs practices and standards to ensure that goods and people entering a country do not pose a threat;
> - Expanding the use of modernized and automated systems, processes, and trade data to make vessel registration, ownership, and operation, as well as

[2]White House (George W. Bush). 2005. *The National Strategy for Maritime Security,* Washington, D.C., September, p. 2.

[3]White House (George W. Bush). 2005. *The National Strategy for Maritime Security,* Washington, D.C., September, p. 3.

[4]White House (George W. Bush). 2005. *The National Strategy for Maritime Security,* Washington, D.C., September, p. 12 and p. 15.

crew and cargo identification, more transparent and readily available in a timely manner;

- Developing, funding, and implementing effective measures for interdicting suspected terrorists or criminals;
- Developing and expanding means for rapid exchanges among governments of relevant intelligence and law enforcement information concerning suspected terrorist or criminal activity in the maritime domain;
- Adopting streamlined procedures to verify nationality and take appropriate and verifiable enforcement action against vessels in a timely manner consistent with the well-established doctrine of exclusive flag state jurisdiction;
- Expanding the U.S. government's abilities to prescreen international cargo bound for the United States prior to lading;
- Adopting procedures for enforcement action against vessels entering or leaving a nation's ports, internal waters, or territorial seas when they are reasonably suspected of carrying terrorists or criminals or supporting a terrorist or criminal endeavor; and
- Adopting streamlined procedures for inspecting vessels reasonably suspected of carrying suspicious cargo and seizing such cargo when it is identified as subject to confiscation.[5]

The NSMS does not alter the existing authorities or responsibilities of U.S. government department and agency heads or the chain of command for military forces.

Interagency Supporting Plans

In conjunction with the development of the NSMS, the departments of the U.S. executive branch developed specific supporting plans. Those most relevant to MSP include the following:

- National Plan to Achieve Maritime Domain Awareness,[6]
- Global Maritime Intelligence Integration (GMII) plan,[7]
- Maritime Operational Threat Response (MOTR) Plan,[8] and
- International Outreach and Coordination Strategy.[9]

---

[5]White House (George W. Bush). 2005. *The National Strategy for Maritime Security,* Washington, D.C., September, p. 15.

[6]Department of Homeland Security. 2005. *National Plan to Achieve Maritime Domain Awareness for the National Strategy for Maritime Security,* Washington, D.C., October.

[7]See <http://www.whitehouse.gov/homeland/maritime-security.html>. Accessed September 26, 2007.

[8]See <http://www.whitehouse.gov/homeland/maritime-security.html>. Accessed September 26, 2007.

[9]U.S. Department of State (Condoleeza Rice). 2005. *International Outreach and Coordination Strategy for the National Strategy for Maritime Security,* Washington, D.C., November.

Of the lead organizations involved in MSP, the Department of Defense (DOD) and the U.S. Coast Guard (USCG) have the largest forces and manage the largest resources. Their operational responsibilities have motivated sophisticated planning, resource allocation, and force development, management, and employment processes.

Current Directives and Guidance

National Security Presidential Directive (NSPD) 41 mandates the "coordination of United States Government maritime security programs and initiatives to achieve a comprehensive and cohesive national effort involving appropriate Federal, State, local, and private sector entities . . . ensuring seamless, coordinated implementation of authorities and responsibilities relating to the security of the Maritime Domain by and among Federal departments and agencies."[10] It established the Maritime Security Policy Coordinating Committee (MSPCC) to "review existing interagency practices, coordination, and execution of U.S. policies and strategies relating to maritime security, and recommend specific improvements to all of them as warranted." It states that the "MSPCC, in consultation with the relevant regional and functional policy coordinating committees of the federal government, and without exercising operational oversight, shall act as the primary forum for interagency coordination of the implementation of this directive."

NSPD 41 also directed the secretaries of Defense and Homeland Security to draft the NSMS, which was promulgated by the President in September 2005,[11] and to prepare supporting plans. Neither NSPD 41 nor the NSMS alters existing authorities or responsibilities of the department and agency heads to carry out operational activities or to provide or receive information.

The agencies involved are responsible for conducting their individual operations to implement the policies in the national strategy and plans. Implementing the visions represented in the NSMS and the supporting plans is a complex undertaking. NSPD 41 identifies the scope of participation to address domestic, international, public, and private components.

The NSMS identifies the threats to maritime security as follows:

- Countries that "provide safe havens for criminals and terrorists, who use these countries as bases of operations to export illicit activities into the maritime

[10]National Security Presidential Directive NSPD-41/Homeland Security Policy Directive HSPD-13, December 21, 2004. Available at <http://www.fas.org/irp/offdocs/nspd/nspd41.pdf>. Accessed June 26, 2007.

[11]White House (George W. Bush). 2005. *The National Strategy for Maritime Security,* Washington, D.C., September. See <http://www.dhs.gov/xlibrary/assets/HSPD13_MaritimeSecurityStrategy.pdf>. Accessed June 26, 2007.

domain and into other areas of the globe."[12] For the purposes of this study, this includes states that lack the ability to enforce national and international laws in areas over which they have jurisdiction (recognizing that all criminal activities occur in all states).

- Proliferation of weapons of mass destruction (WMD) and their means of delivery, with particular concern that those weapons will become available to organizations that use terrorism to pursue their objectives.
- Terrorist attacks from or in the maritime domain (including ports and offshore facilities) or that use the maritime domain to foster and support their activities.
- Cyberattacks on information systems that are integral to maritime operations.
- Criminal activities, including smuggling people, drugs, weapons, and other contraband, as well as piracy and armed robbery against vessels, particularly in the pay of terrorists and in regions where there is little or no maritime law enforcement capacity.
- Environmental destruction and management of maritime resources that contribute to aggressive actions.
- Illegal seaborne migration, which tactic also may be used by terrorists to enter a target country.

The NSMS also calls for minimizing damage and expediting recovery in the case of a natural disaster such as Hurricane Katrina and assisting the partners to maintain sovereignty of the seas over which they have jurisdiction.

Finding: Extensive coordination among essentially all U.S. government agencies is required to implement the National Strategy for Maritime Security and associated plans. However, the committee found little evidence of any broad coordination of activities by these agencies following the introduction of the NSMS.

U.S. PARTICIPATION IN MSP

The NSMS provides a basis for bringing together all of the federal government's relevant departments and agencies in order to meet the maritime security challenges described above. It can also provide a framework for coordinating all maritime security initiatives with foreign governments and international organizations as well as soliciting international support for enhanced maritime security. Under the MSP concept, the United States will be able to work with its partners in developing regional maritime security capabilities based on the needs and expectations of countries in various regions of the world. The MSP concept provides

[12]White House (George W. Bush). 2005. *The National Strategy for Maritime Security,* Washington, D.C., September, p. 3.

regional maritime security frameworks based principally on bilateral agreements and consistent with international law and United Nations conventions.

The Spectrum of Maritime Security and the U.S. Navy

Maritime security for the Navy today has evolved from conventional maritime operations against a peer competitor to dealing with an environment rife with asymmetric threats and supporting law enforcement functions in the maritime domain. The recognition that nations have common interests in maritime security and can work together to develop peaceful change has led to the Chief of Naval Operations' (CNO's) concept of the "1,000-ship Navy," whereby the United States enters into some form of maritime partnership with willing seafaring nations across the world. Only by working together can countries protect their interests in the maritime domain from the complex challenges they face today. Although this idea represents a cultural change from its classical warfighting missions, the Navy has a rich tradition of operating in green and brown waters all over the world. The variety of Navy missions with the potential to achieve maritime security is illustrated in Figure 4.1.

The Navy, the USCG, and Law Enforcement

Increasingly the Navy is involved in a variety of joint operations with the USCG and law enforcement agencies of the U.S. government. In many instances the Navy finds itself supporting the USCG units as well as other agencies, because the USCG has better access in many parts of the world and possesses law enforcement authorities. White hulls are accepted where gray hulls are not in some parts of the maritime domain, allowing the USCG to take the lead in a variety of initiatives in support of maritime security requirements. The employment of USCG units in conjunction with other law enforcement agencies and foreign maritime partners can provide a significant capability in some maritime environments. For the USCG the spectrum of maritime security activities will provide operational challenges across the law enforcement domain (see Figure 4.1).

The International Impact of MSP

MSP is an international association of maritime nations that participate in international commerce and have a stake in security and freedom of the seas. Such partnerships are necessary in today's world to confront the complex shared challenges and to maintain stability. Partners in the maritime domain would assist all countries in using the sea for lawful purposes as well as legitimate commerce. A partnership would not be led by any one country and membership would be voluntary, with the goal of building partner capacity through shared maritime security, situational awareness, and information.

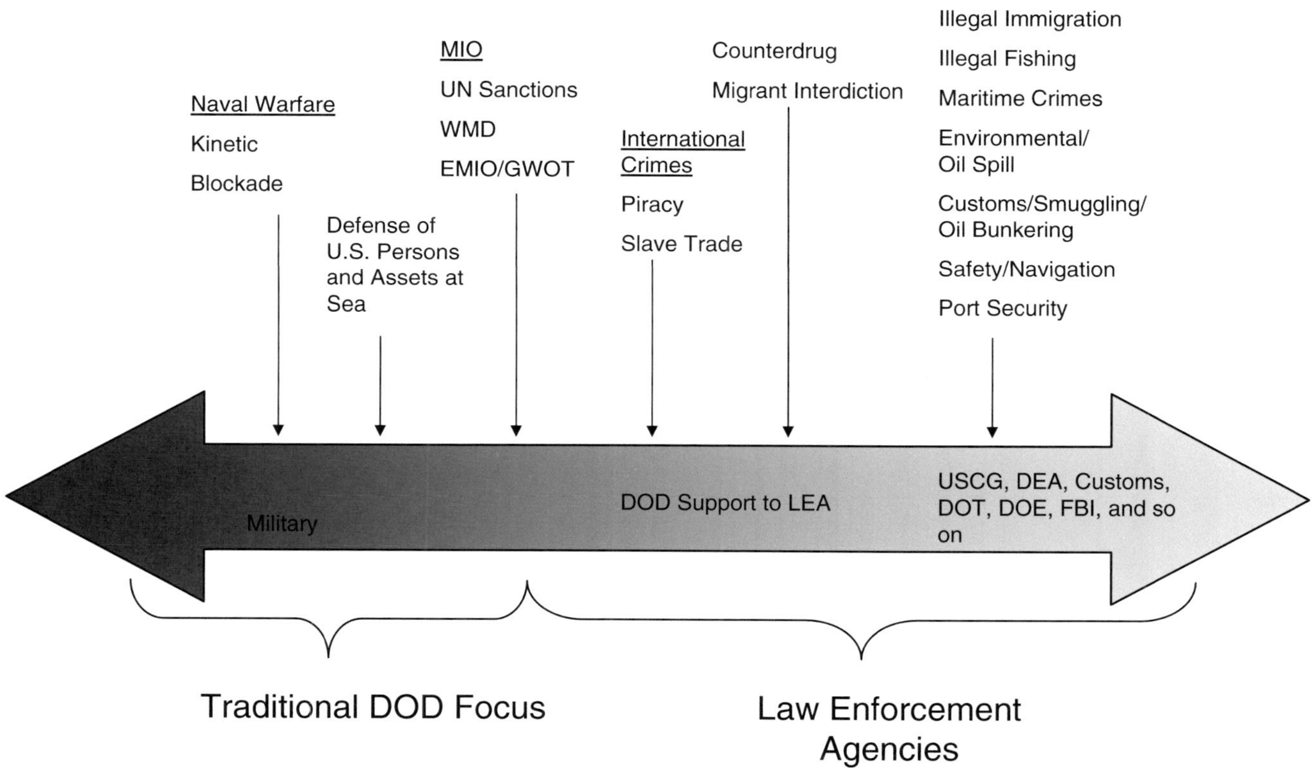

FIGURE 4.1 Spectrum of activities for maritime security. NOTE: DEA, Drug Enforcement Agency; DOE, Department of Energy; FBI, Federal Bureau of Investigation; EMIO, Expanded Maritime Intercept Operation. For additional definitions, see Appendix G.

Partners should recognize and support the vital role of international organizations that engage in maritime security and law enforcement issues. As well, maritime nations will more readily accept the United States as a partner in maintaining free and open use of the maritime domain when the U.S. Senate ratifies the United Nations Convention on the Law of the Sea (UNCLOS). This topic is discussed in greater detail in Appendix C. Our nation's stature with organizations such as the International Maritime Organization (IMO), the International Labor Organization (ILO), Interpol, and Lloyd's would also be enhanced. The World Meteorological Organization with its 180 member nations is another place where there are already international agreements in place for navigation, hydrographic surveys, and foreign student educational exchange programs. In the private sector, nongovernmental organizations, shipping companies, and other commercial assets could all support the role of the international community. If the vision of MSP is to be realized, the efforts of many will have to be combined.

The International Reaction to the "1,000-Ship Navy" Idea

The concept of a 1,000-ship Navy gained widespread attention from the attendees at the 2006 International Sea Power Symposium hosted by the CNO.[13] Since then, leaders of maritime forces from around the world have reacted favorably to this concept and have crafted their comments based on a regional perspective as well as on the contribution that MSP can achieve in the maritime commons. The comments tend to follow certain themes depending on the particular challenges that confront naval leaders today. Many of the leaders indicate that terrorism, the proliferation of WMD, transnational criminal and piracy threats, globalization, competition for resources, demographic shifts, and the impact of climate change are all concerns that they face. They generally support appropriate information exchange and the ability to work more closely in peacekeeping and stability operations while maintaining the capability to respond to regional challenges as they arise. The need to respect national characteristics and cultures as well as regional desires was commented on. The point was made that the sea cannot be commanded. Also, there needs to be an interagency approach to maritime security at the international level to get the proper support for elements operating in the regional maritime domains.

While many world naval leaders have expressed support for the 1,000-ship Navy, there is no assurance that their governments are committed to active participation. Personal relationships at the diplomatic, military, and law enforcement levels are essential to building trust. Knowledgeable and trusted foreign area officers (FAOs) will prove invaluable in convincing regional navies that they must work to guide their countries toward participation in the now regional and later

[13]Chief of Naval Operations (ADM Michael G. Mullen, USN) in remarks delivered at the 17th International Seapower Symposium, Naval War College, Newport, R.I., September 21, 2005.

global sharing of maritime information. Therefore, it seems prudent for the CNO, CMC, and CCG to ensure that the cadre of service foreign area officers becomes expert on the governance of the maritime domain.

DOD and DHS Force Planning for MSP

DOD planning to support operations for MSP would start when requests for assistance are received from the combatant commanders (COCOMs), through their naval component commanders, to support their theater engagement plans. DOD, as well as the Department of Homeland Security (DHS) and the USCG, would also provide assistance to the State Department and the various federal agencies and departments, as appropriate, in support of the national maritime strategy. Navy assets are routinely provided to the COCOMs in response to requests for contingency planning, exercises with allies, and forward presence deployments. Ships that are out on normal deployments could be tasked to support maritime security missions, respond to humanitarian disasters, or take part in stability operations in the littorals. Any of these could be carried out in response to a request for assistance from the U.S. ambassador in a given country. USCG assets could also be assigned similar missions. One of the best examples of this would be the deployment of one USCG cutter to the Gulf of Guinea for an extended period of time to support the U.S. European Command (EUCOM) activities in that region. The key questions are these: What assets are available to carry out the tasking? What capabilities are needed? Are there ships available with the necessary equipment? An available Aegis cruiser, for example, has a tremendous warfighting capability but may be totally unsuited for that specific mission. The need for advance planning for such missions is obvious if DOD and the COCOMs want to maximize the impact of their maritime security operations.

Optional Capabilities for Maritime Security Operations

In addition to normal deployments, where ships may be tasked to carry out specific missions, other emerging deployment concepts could fit into the MSP effectively. Such deployments could be coordinated through the naval component commander in the U.S. Pacific Command (PACOM), the U.S. Southern Command (SOUTHCOM), EUCOM, and the U.S. Central Command (CENTCOM).

Hospital Ships

Two U.S. hospital ships had a tremendous impact in the regions where they were recently deployed. Such deployments improve relationships with the countries where the visits take place and build trust. The USNS *Mercy,* which deployed to Southeast Asia and the Indian Ocean in 2006, had a mixed crew made up of U.S Navy and security personnel. Also part of the crew were members of nongovernmental organizations (NGOs), U.S. medical staff, and medical teams from other countries trained to carry out medical and dental tasks in the host country. This was a first time for many NGOs onboard a U.S. ship. The results of this effort were extremely positive in the countries where the *Mercy* visited. The second deployment was that of the USNS *Comfort,* which operated in countries around the Caribbean and South America for 4 months during 2007. The crew consisted of Navy, Air Force, and Air Guard personnel, as well as NGOs, Public Health Service specialists, a band, and linguists. Of note here is the extensive planning before each deployment by the Navy, the other federal agencies, the COCOMs, and the country teams in each country that hosted the visit. The success of the hospital ship deployments could be repeated by assigning a similar mission to hospital-configured amphibious ships such as the LHA (amphibious assault ship, general purpose), the LHD (amphibious assault ship, multipurpose), and the LPD (amphibious transport dock). The ability to conduct Phase Zero Stability Operations with these kinds of assets will do a lot to strengthen relations with other countries, but they have to be carefully planned and coordinated to suit the regions where they will operate.

The Global Fleet Station Concept

This new CNO initiative, still under development, has the potential for providing excellent service in support of partnership and enabling activities in different parts of the world. Global Fleet Station is a persistent sea base of operations from which to coordinate and employ adaptive force packages in an area of interest. Global Fleet Station offers a means to improve regional maritime security through bilateral and multilateral cooperative efforts and efforts with NGOs. Two early applications of this concept have had positive results. The first deployment was to the Caribbean in support of SOUTHCOM requirements and to try out the concept. The results were very encouraging. At the same time the USCG had a support tender on station in the Caribbean that carried out a variety of maintenance tasks in support of host countries in the region. The second deployment under this concept will be a landing ship dock that will go to the Gulf of Guinea to replace a USCG cutter for an extended period of time. This deployment has been planned carefully by host nations in the region, EUCOM, the various agencies, the State Department, and the Navy. It will have another mixed crew comprising Navy personnel, training teams, medical personnel, and other experts who will work in various countries in the region. These kinds of deployments will play a significant role in support of regional initiatives within the MSP.

SHORTFALLS IN OPERATIONAL FUNCTIONS

NSPD 41 and the associated strategies and plans address only policy formulation and coordination, but MSP also involves operational functions:

- Strategic and operational planning;
- Resource allocation;
- Development, management, and employment of the military and law enforcement forces and other capabilities needed to provide maritime security;[14] and
- Performance assessment and feedback.

Table 4.1 indicates the organizational leads for these functions, where they have been clearly identified, for each activity of the MSP. As the table shows, responsibilities for setting the policies for the various activities to be conducted under MSP are spread widely across the U.S. government. Moreover, no agency has been designated to conduct strategic and operational planning for MSP, to identify the resources needed and develop the capabilities to implement them, or to conduct the force management to schedule and employ the military and law enforcement forces involved.

Finding: Major gaps in roles and responsibilities exist between, on the one hand, the agencies with responsibilities and authorities for setting policy and establishing regulations and, on the other hand, the maritime forces responsible for enforcing these regulations.

The current roles and responsibilities for various maritime security activities include the following:

- *Countering the proliferation of WMD.* A separate Policy Coordinating Committee is responsible for policy on the proliferation of WMD. Also, within the DOD, the commander of the U.S. Strategic Command is assigned principal responsibility for this mission by the President in the Unified Command Plan; the Special Forces Command has the forces trained for sophisticated operations to recover WMD, and the regional COCOMs are allocated the naval and, occasionally, the USCG forces that would be involved in interdicting WMD at sea.
- *Countering terrorism.* The National Counter Terrorism Center is responsible for "leading the USG [U.S. government] in Counterterrorism Intelligence and Strategic Operational Planning in order to combat the terrorist threat to the US and its interests."[15]
- *Countering cyberattacks.* Like the maritime domain, the vast majority of cyberspace is privately owned. DHS is leading U.S. government efforts to secure

---

[14]Force development, management, and employment are used by DOD to describe the capabilities needed for the forces to conduct assigned missions from current to anticipated missions decades in the future, the management of current forces over the next several years with respect to personnel and unit rotation policies, and the actual employment of forces in operations and training for future operations.

[15]Mission statement of the National Counterterrorism Center, see <http://www.nctc.gov/>. Accessed August 29, 2007.

TABLE 4.1 Maritime Security Partnership Functional Responsibilities by Activity

	Function				
Activity	Policy/ Guidance	Strategic and Operational Planning	Resource Allocation	Capability Development	Force Management and Employment
Countering proliferation of WMD[a]	Counter-proliferation PCC				
Countering terrorism[b]		National Counterterrorism Center[c]			
Countering cyberattacks	DHS				
Smuggling, piracy, armed robbery at sea	DoS				
Countering environmental destruction	EPA, DoS				
Fisheries protection	Regional fisheries management boards		NOAA		
Preventing illegal seaborne migration	Immigration and customs enforcement				
Humanitarian assistance and disaster relief	Foreign: DoS, DOD Domestic: FEMA		DoS, DOD		
Building partnership capacity	DoS	DOD[d]	DoS, DOD		
Pandemic mitigation	HHS				

NOTE: Gaps in column entries indicate that, as of this writing, no agency had been designated as having responsibility for the function indicated. PCC, Policy Coordinating Committee; DHS, Department of Homeland Security; DoS, Department of State; EPA, Environmental Protection Agency; FEMA, Federal Emergency Management Agency; HHS, Department of Health and Human Services; NOAA, National Oceanic and Atmospheric Administration.

[a]The guidance includes White House (George W. Bush), 2002, *National Strategy to Combat Weapons of Mass Destruction* (unclassified version), NSPD-17/HSPD 4, December. Available at <http://www.fas.org/irp/offdocs/nspd/nspd-17.html>. Accessed August 29, 2007.

[b]The guidance includes White House (George W. Bush), 2006, *National Strategy for Combating Terrorism,* September. Available at <http://www.fas.org/irp/offdocs/nspd/nspd-17.html>. Accessed August 29, 2007.

[c]The Intelligence Reform and Terrorism Prevention Act (IRTPA) codified the responsibilities of the National Counterterrorism Center in 2004. See <http://www.fas.org/irp/congress/2004_rpt/h108-796.html>. Accessed August 28, 2007.

[d]DOD theater security cooperation plans are coordinated with the Department of State (DoS), with the DoS having the principal responsibility and authority for security assistance. Section 1206 of the Fiscal Year National Defense Authorization Act authorized DOD to expend funds directly to assist foreign military forces in countering terrorism.

critical infrastructure, monitor the health of cyberspace, and respond to major incidents and attacks.[16]

- *Countering smuggling, piracy, and armed robbery at sea.* The Department of State (DoS), working with the USCG, represents the United States in the IMO and in developing agreements with other nations regarding responsibilities and authorities for countering smuggling, piracy, and armed robbery at sea (see Chapter 3). Most of the effort to counter smuggling has been focused on interdicting illegal drugs. Appendix C discusses joint interagency task forces that have been created to address this problem. NSPD 22 established a Cabinet-level Interagency Task Force to Monitor and Combat Trafficking in Persons, which involves the DoS and the Departments of Justice (DOJ) and Labor (DoL).[17]
- *Countering environmental destruction.* The Environmental Protection Agency (EPA) sets regulations to implement U.S. environmental law; some of the regulations directly relate to coastal zones, marine protection, and ocean dumping.[18] The DoS, with strong USCG participation, leads U.S. interactions with other nations to harmonize national and international environmental laws and conventions, including the international conventions for
  —The Prevention of Pollution from Ships, 1973/1978;
  —Intervention on the High Seas in Cases of Oil Pollution Casualties, 1969;
  —Prevention of Maritime Pollution by Dumping of Wastes and Other Matter, 1972;
  —Oil Pollution Preparedness, Response, and Cooperation, 1990;
  —Preparedness, Response, and Co-operation to Pollution Incidents by Hazardous and Noxious Substances, 2000;
  —Control of Harmful Anti-fouling Systems on Ships, 2001; and
  —Control and Management of Ship's Ballast Water and Sediments, 2004.
- *Fisheries protection.* U.S. Code Title 50, Chapter VI, specifies the procedures for fishery conservation and management and associated responsibilities of the Department of Commerce and NOAA.[19] The law provides for regional fisheries management councils that prepare statements of organization, practices, and procedures and are funded by federal grants. Enforcing these procedures falls principally to the USCG. The decentralized approach, which extends from the

---

[16] The guidance includes White House (George W. Bush), 2003, *National Strategy to Secure Cyberspace*, February. Available at <http://www.whitehouse.gov/pcipb/cyberspace_strategy.pdf>. Accessed August 28, 2007.

[17] See <http://www.fas.org/irp/offdocs/nspd/trafpers.html>. Accessed August 28, 2007.

[18] See <http://www.epa.gov/epahome/lawintro.htm>. Accessed August 28, 2007.

[19] See <http://ecfr.gpoaccess.gov/cgi/t/text/textidx?c=ecfr&sid=67c522ccd6dd3464c7455787c234c21a&rgn=div8&view=text&node=50:8.0.1.1.1.2.1.5&idno=50>. Accessed June 27, 2007.

United States into the international arena, presents many challenges to effective fisheries management.[20]

• *Preventing illegal seaborne migration.* The U.S. Immigration and Customs Enforcement (ICE) agency within the DHS has primary responsibilities for preventing illegal migration across U.S. land and sea borders:

> ICE investigates a wide range of national security, financial and smuggling violations including drug smuggling, human trafficking, illegal arms exports, financial crimes, commercial fraud, human smuggling, document fraud, money laundering, child pornography/exploitation and immigration fraud.[21]

• *Humanitarian assistance and disaster relief.* Congress appropriates

> . . . overseas Humanitarian, Disaster, and Civic Aid (OHDACA) funds to augment combatant commander capabilities to respond rapidly and effectively to humanitarian crises, thereby allowing U.S. military forces to obtain substantial training and access benefits by participating in OHDACA activities enhancing readiness across a number of operational areas—including C3I [command, control, communications and intelligence], civil affairs, civil and combat engineering, explosive ordnance disposal, logistics, medical, and special operations.[22]

Combatant commanders allocate these funds in close coordination with the U.S. ambassador of the affected country. The Federal Emergency Management Agency (FEMA), within the DHS, is responsible for federal assistance in the event of domestic disasters.

• *Building partnership capacity.* Interest in and efforts to build partnership capacity have grown with experiences in operations in Iraq, Afghanistan, and the global war on terrorism. DOD published the directive "Military Support for Stability, Security, Transition, and Reconstruction (SSTR) Operations"[23] in November 2005 and developed a roadmap for building partnership capacity in conjunction with the 2006 Quadrennial Defense Review (QDR). The roadmap outlines options to improve the collective capabilities and performance of DOD and its partners at home and abroad. It identifies ways to enhance international unity of effort by improving the capacity and capability of international partners and international cooperation on homeland defense matters.[24] The Under Secretary of Defense for Policy established a new office (Assistant Secretary of

---

[20]Patricia Lee Devaney, "Regional Fisheries Management Organizations: Bringing Order to Disorder." Available at <http://www.pon.org/downloads/ien14_4Devaney.pdf>. Accessed August 29, 2007.

[21]See <http://www.ice.gov/about/faq.htm>. Accessed July 1, 2007.

[22]See <http://www.dsca.osd.mil/programs/HA/OVERSEAS%20HUMANITARIAN%20DISASTER%20AND%20CIVIC%20AID.pdf>. Accessed July 1, 2007.

[23]Department of Defense Directive 3000.05, "Military Support for Stability, Security, Transition, and Reconstruction (SSTR) Operations," November 28, 2005. Available at <http://www.fas.org/irp/doddir/dod/d3000_05.pdf>. Accessed July 28, 2007.

[24]Deputy Secretary of Defense (Gordon England). 2007. *Second Quarterly Report to Congress on Implementation of the 2006 Quadrennial Defense Review,* April 30, p. 2.

Defense for Global Security Affairs and Deputy Assistant Secretary for Building Partnership Capacity) charged with providing focus on the security cooperation assessment process. The 2006 National Defense Authorization Act, Section 1206, authorized funding for DOD to train and equip foreign military forces to conduct counterterrorism and stability operations. This effort led the DOD to propose an act on building global partnership.[25] The Office of Management and Budget (OMB) requested that the Senate include support for local populations related to humanitarian relief and reconstruction in FY07 in its appropriations to extend the activities related to Section 1206.[26] The proposed National Defense Authorization Act for Fiscal Year 2008, Section 1202, "extends for one year the authority granted to the Department by Section 1207 of the FY06 Defense Authorization Act to provide the Secretary of State with services, defense articles, or funding to facilitate the State Department's efforts to provide reconstruction, security, or stabilization assistance to a foreign country." This provision increases the aggregate amount of support that may be provided by the DOD to the DoS in FY08 to $200 million. Section 1207 authority may, among other things, be used to support DoS programs and authorities to train and equip foreign police, gendarmerie, constabulary, and internal defense forces to enhance security and stability. This authority differs from, but complements, the authority granted by Section 1206 of the FY06 Defense Authorization Act, which authorizes the Secretary of Defense (with the concurrence of the Secretary of State) to build the capacity of a foreign nation's military forces in order for that nation to conduct counterterrorist operations and to participate in or support military and stability operations in which the United States is a participant. So-called Section 1206 authority remains authorized at the level of $300 million for FY08.[27] The proposed Building Global Partnership Act would permanently authorize such activities. Portions of the proposed act related to increased funding for OHDACA and a permanent global Commander's Emergency Response Program (CERP), currently authorized only for operations in Iraq and Afghanistan, were removed and are being coordinated with Congress.[28] The Defense Security Cooperation Agency within the DOD was established to better coordinate the ability of DOD and the Department of State to provide security assistance across the wide variety of programs that exist.[29]

---

[25]Deputy Secretary of Defense (Gordon England). 2007. *Second Quarterly Report to Congress on Implementation of the 2006 Quadrennial Defense Review,* April 30, Appendix 3, p. 4.

[26]See <http://www.whitehouse.gov/omb/legislative/sap/109-2/s2766sap-s.pdf>. Accessed July 28, 2007.

[27]See <http://rpc.senate.gov/_files/L260DefAuthS1547070907MS.pdf>. Accessed July 28, 2007.

[28]Deputy Secretary of Defense (Gordon England). 2007. *Second Quarterly Report to Congress on Implementation of the 2006 Quadrennial Defense Review,* April 30, Appendix 3, p. 6.

[29]The Defense Security Cooperation Agency (DSCA) administers the Foreign Military Sales program and the associated Foreign Military Financing program as well as the International Military Education and Training program, which mostly brings foreign military students to schools in the United States but also finances some mobile training teams to train in the countries themselves. The Services contract for equipment and other services with U.S. companies. Since the law mandates that the business be conducted on a no-profit/no-loss basis, DSCA charges the country customers 3 percent

Coordinating such assistance across the involved agencies to address strategic objectives for even one foreign country remains a daunting task.

Of the lead organizations involved in MSP, the DOD and the USCG have the largest forces and manage the largest resources. Their operational responsibilities have motivated sophisticated planning, resource allocation, and force development, management, and employment processes.

FOUNDATIONS OF MARITIME SECURITY PARTNERSHIPS

The DOD's theater security cooperation (TSC) plans address the set of operational functions needed for DOD's participation in MSP.

The concepts and implementation of TSC have evolved over the last decade. With the end of the Cold War, government and academic institutions sought to understand and adapt to the new security environment. At the National Defense University Pacific Symposium in 1991, the Assistant Secretary of Defense for International Security Affairs introduced the notion of cooperative vigilance as one approach to Asia-Pacific security.[30] The notion was an adaptation of cooperative vigilance among animal herds and flocks of birds, whereby members alternate their watch duties for the group. A conference of some of the nation's most prominent foreign policy and arms control scholars at Stanford University in April 1992 proposed that cooperative engagement to achieve multinational security should replace Cold War concepts of national security.[31] ADM Charles R. Larson, USN, the Commander in Chief of PACOM, began institutionalizing cooperative engagement within his command. This approach applied "military assets, funds, and programs to achieve three objectives: forward presence, strong alliances, and crisis response. . . . The forward deployment of the U.S. forces in the region contributes significantly to maintaining stability, enhances our diplomatic influence, and promotes an environment conducive to the growth of our economic interests there." The intent was to "seize the opportunity offered in this new era to shape a better world—one built on shared ideas, interests, and responsibilities" and "engenders [the building of] coalitions for collective action in time of crisis."[32]

This approach became a national strategy with the publication of the 1996

---

on each sale (even if financed by the United States) to cover the Services' costs of administering each case, as well as DSCA's own costs. DSCA and its Defense Cooperation Offices in the countries need to urge the countries, the U.S. Navy, the selling American companies, and the COCOMs to make sure the equipment and information capabilities sold to the countries are interoperable with United States and other international systems.

[30]See Seng Tan and Amitay Acharya. 2004. *Asia-Pacific Security Cooperation: National Interests and Regional Order,* M.E. Sharpe, Armonk, N.Y., p. 227.

[31]See <http://news-service.stanford.edu/pr/92/920408Arc2318.html>. Accessed August 28, 2007.

[32]Charles Larson. 1993. "Cooperative Engagement," *Joint Force Quarterly*, Issue 2 (Autumn), p. 82.

national security strategy "Engagement and Enlargement."[33] In conjunction with this strategy, the Chairman of the Joint Chiefs and the COCOMs began formalizing their theater engagement plans. In 2001, the new administration put a somewhat different emphasis on this effort, changing "theater engagement plans" to "theater security cooperation plans," which initially emphasized the more traditional aspects of international military-to-military interactions. However, the global war on terrorism restored the emphasis on interactions involving nontraditional security challenges and the perceived value of humanitarian assistance in promoting the U.S. image and values.

The COCOMs prepare TSC plans to carry out the missions assigned in the Security Cooperation Guidance provided by the Secretary of Defense. According to the 2006 Quadrennial Defense Review, the Security Cooperation Guidance will be incorporated into the Contingency Planning Guidance, signed by the President, which directs COCOM planning. As with the previous theater engagement plans, TSC plans are carefully coordinated with mission performance plans of the U.S. ambassador in each country and between the policy organizations within DOD and DoS. Each subordinate military service and functional component commander recommends to the COCOM which interactions with foreign nations, from leader and ship visits, to medical, dental, and other humanitarian visits, to major military exercises, should be included in the theater plan. In practice, the Navy component commanders (the four-star area fleet commanders) have received approval for their proposed uses of assigned and allocated Navy forces approved within the TSC plans. Such plans guide DOD's authorized activities in connection with MSP, including the new emphasis on working with the maritime industries.

Obtaining U.S. Forces for MSP Activities

The main thrust of this report is to support the establishment of mechanisms by which the U.S. government in general and the U.S. Navy in particular can help other nations improve their own maritime security situation. The vehicle for this support is the greatly increased collection and distribution of MDA information. Responsibility for an adequate response to emerging threats to U.S. maritime interests rests primarily with the naval forces. This section addresses the impact of the MSP initiative on U.S. naval force planning.

By law, all military combat forces must be assigned to a COCOM. Naval forces are assigned principally to U.S. Joint Forces Command and PACOM. Though forces are assigned to COCOMs, they are also apportioned for major contingency plans and allocated to COCOMs to conduct exercises and operations and to respond to humanitarian crises as required.[34]

---

[33]See <http://www.fas.org/spp/military/docops/national/1996stra.htm>. Accessed July 28, 2007.

[34]Adaptive Planning Overview, see <http://www.mors.org/meetings/cbp/presentations/Hoffman-Mon.pdf>. Accessed August 5, 2007.

COCOMs request forces to support joint training and TSC plans. The U.S. Joint Forces Command recommends the forces to be allocated for the proposed deployments and exercises, the Global Force Management Board reviews and coordinates these recommendations, the Chairman of the Joint Chiefs of Staff then recommends the forces to be deployed to the Secretary of Defense, who then signs deployment orders for those forces not assigned to the COCOM. COCOMs can deploy assigned forces within their area of operations without the Secretary of Defense signing deployment orders.

To support adaptive planning and global force management, the DOD has issued an instruction that "establishes policy and assigns responsibility under Reference (a) [Strategic Planning Guidance (SPG) FY 2006-2011, March 1, 2004] for developing standardized force structure data that will provide on-demand information in a net-centric environment."[35] As agreed with the DHS, the USCG will make data on the readiness of its forces available to support global force management.

Finding: COCOM Theater Security Cooperation plans provide the foundation for conducting the operational functions that are not incorporated into NSPD 41 or its supporting plans. However, no similar procedures exist across the other government agencies that have authority or responsibilities for MSP activities that go beyond the DOD.

U.S. Navy Role

The concept originally called the 1,000-ship Navy has gained support from some 24 other chiefs of navies, a sign that the U.S. Navy can lead the U.S. participation in this maritime security effort. However, the CNO's concept of the 1,000-ship Navy addresses a litany of problems that beg for solutions. The problems go well beyond the interests of the Navy—in fact, they affect every cabinet-level department in the U.S. government. MSP must be an international initiative whereby countries participate on the basis of their national interests as well as regional policy agreements and maritime law. There is currently no single agency or department that can effectively speak for the President and the nation's maritime concerns. Responsibilities are fragmented. Authority is often exercised but decisions are not coordinated, so the result is less than optimal. If there is to be a professional, comprehensive, internationally respected entity of the U.S. government dealing with maritime affairs, it needs to be able to undertake many responsibilities:

- Foster maritime commerce,

[35]DODI 8260.03, "Organizational and Force Structure Construct (OFSC) for Global Force Management (GFM)," August 23, 2006, p. 1. Available at <http://www.dtic.mil/whs/directives/corres/html/826003.htm>. Accessed August 5, 2007.

- Represent the United States to all international and regional maritime organizations,
- Establish maritime agreements on behalf of the United States,
- Represent the U.S. interest in all matters pertaining to the UNCLOS and maritime law enforcement,
- Develop and field systems to give the United States effective MDA,
- Formulate top-level policy for the establishment and sustainment of maritime aids to navigation,
- Set policy for maritime traffic rules and systems,
- Establish and maintain a uniform national policy for U.S. access to other countries' ports and for U.S. port security,
- Provide policy oversight for the safe operation and security of U.S. flag vessels,
- Establish and maintain uniform standards for training and certification of mariners,
- Provide oversight of standards for maritime vessel construction, and
- Enforce maritime environmental standards.

The policy pronounced by the President in December 2004 for the implementation of a comprehensive NSMS called for the creation of the MSPCC at the National Security Council (NSC).

Finding: The MSPCC and its parent, the NSC, have not yet met the requirements of the 2004 mandate, as the discussion above indicates, nor have they developed even short-term initiatives to give the nation a robust capability for MDA. A new, invigorated approach must be undertaken to meet national maritime needs.

STRATEGIC INTERACTION WITH INTERAGENCY INITIATIVES

Interagency support for U.S. participation in the MSP is crucial if the concept is to work; however, the support has been sporadic so far. Initiatives within the NSC and DoS are being coordinated but have to be resolved if the Navy, the USCG, and law enforcement agencies are to be effective when operating in the maritime domain in support of the NSMS requirements in different regions of the world. Several interagency initiatives and programs are part of this effort, but they work independently rather than together at the moment. Several of the more important interagency initiatives present obvious coordination challenges (also see Figure 4.2):

- *Proliferation Security Initiative (PSI).* This initiative seeks to stop the shipments of WMD to and from states and nonstate actors worldwide. Seventy countries have indicated support for PSI, while 20 are actively participating in this effort.

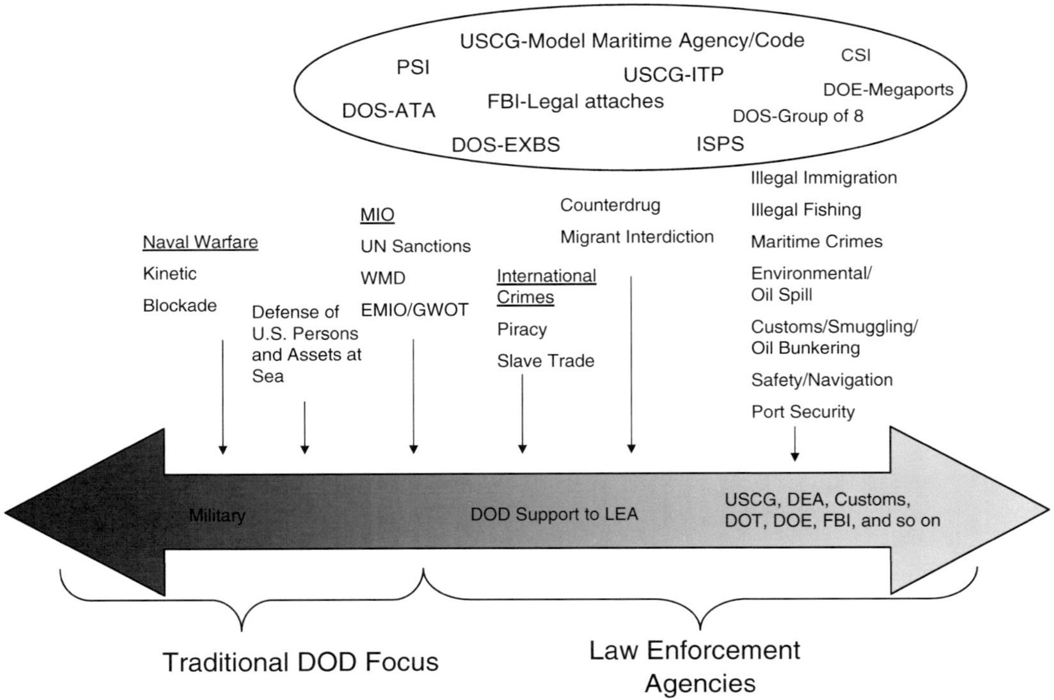

FIGURE 4.2 Supporting interagency initiatives. NOTE: PSI, Proliferation Security Initiative; CSI, Container Security Initiative; ITP, International Training Programs (USCG); ATA, Antiterrorism Assistance; EXBS, Export Control and Border Security; ISPS, International Shipping and Port Security; EMIO, Expanded Maritime Intercept Operation; GWOT, global war on terror; DOT, Department of Transportation.

- *Antiterrorism Assistance (ATA).* This initiative is located in the Diplomatic Security Services Training Directorate of DoS. ATA provides training and equipment based on onsite needs assessments for foreign law enforcement and civilian security organizations.
- *Regional Maritime Security Program (RMSP).* This program is jointly coordinated by PACOM and DoS. It is a capacity-building program that is focusing on enhancing cooperative security and maritime law enforcement capabilities in the East Asia and Pacific regions.
- *Export Control and Border Security (EXBS) assistance.* This initiative is a key tool in stemming the proliferation of WMD and related weapons and technologies. It works to ensure that the manufacturers and suppliers have proper control over the export of munitions, dual-use goods, and related technologies. It also tries to ensure that transit and transshipment countries have the tools to interdict illicit shipments across their territories.
- *Group of Eight.* The G-8 Lyon-Roma Group has devised methodology and a checklist for auditing port and maritime security. The procedure has been adopted by the IMO as an international self-assessment checklist.
- *Model Maritime Agency/Code.* This model maritime service code can identify the legal authority that a multimission maritime service needs to function effectively. Developed by the USCG, it has been presented to over 20 countries.
- *International Training Program (ITP).* Provides training programs at USCG schools as well as mobile training teams for members of the international maritime community.
- *Container Security Initiative (CSI).* This initiative proposes a security regime to ensure that all containers that pose a risk for terrorism are identified and inspected at foreign ports before they are loaded on vessels destined for the United States.
- *Organization of American States (OAS) port security assistance.* U.S. missions to the OAS and working with the OAS, the Inter-American Committee Against Terrorism (CICTE), the Inter-American Committee on Ports (CIP), and the Maritime Administration (MARAD) offer assistance to OAS member states to enhance security at their ports in order to comply with the dictates of the IMO.
- *Megaports Initiative.* This initiative helps countries with major international ports to enhance their ability to screen cargo at those ports. It also works to improve radiation detection equipment as well as train personnel in the use of such equipment.
- *FBI legal attachés.* Legal personnel are located in over 50 key cities worldwide and are providing coverage for over 200 countries, territories, and islands. Each office is established through an agreement with the host nation and is normally located in the U.S. embassy in that nation.
- *International Ship and Port Facility Security (ISPS) Code.* ISPS is another IMO initiative implemented in the United States by the USCG, which encourages bilateral or multilateral discussions with other nations to exchange information

on enforcement requirements for international maritime security standards. The USCG works closely with U.S. trade partners to promote reasonable implementation and enforcement of the ISPS Code for enhanced maritime security (see Figure 4.2).

- *Automatic Identification System (AIS).* This is another IMO initiative that allows for ship tracking and monitoring for the Vessel Tracking System (VTS). The International Convention for the Safety of Life at Sea (SOLAS) agreements now require AIS on all ships of 300 GT and over engaged in international voyages as well as all passenger ships regardless of size. The Navy is putting AIS on all of its ships.
- *FAO expertise.* The special skills required of foreign area officers (FAOs) are put to use working with the many programs related to MDA. A career FAO, whether or not he or she comes from one of the sea services, is expected to have assignments in headquarters offices both stateside and abroad. Demonstrated excellence in the leadership of maritime operations should ultimately allow the FAO to be assigned at the three-star level as director of any office or agency tasked with establishing maritime partnerships.

Potential Solutions to the Poor Outlook

A review of the diverse interagency/interdepartment programs, plans, needs, and initiatives described above predicts that it will be very difficult to find a common basis for achieving MSP. Furthermore, a detailed analysis of the requirements to implement any maritime security partnership by the U.S. government turns up a quagmire of bureaucratic and political hurdles that cannot be overcome using traditional organizational tools. This makes it unlikely that the departments and agencies of the U.S. government will be able to execute the President's NSMS.

Several alternatives could be pursued to implement and strengthen MSP both domestically and internationally, among them the following:

- Maintain the current roles and responsibilities for maritime security within the various agencies and departments of the government but improve on interagency coordination mechanisms. Coordination could begin without bureaucratic delay if the NSC would put into effect the already established maritime security coordinating policy. This could be expedited by making the NSC responsible, in accordance with its charter, for an up-to-date report on the implementation status of NSPD 41.
- Assign one agency as the lead for maritime security and increase its role, responsibility, and authority for interagency coordination for maritime security. The lead agency could be the DoS, the USCG within the DHS, or the Navy within the DOD. Because of the broad scope of the maritime problem and the need for consistent national policies, only DoS has the breadth of experience to

be assigned as lead agency; however; it lacks operational resources. The USCG also has broad experience and acceptance and could satisfy many of the needs of the partnerships. The Navy has ownership of the MSP concept but, along with DOD, presents a military front, which may be undesirable. DOD and the Navy also lack law enforcement authority.

- Establish a new agency for maritime security as a standalone agency or within one of the departments. A new agency speaking as a single voice on maritime matters would have a broader mandate on maritime security, including commercial and environmental aspects. It might not, however, fit well under DHS. A new agency would have the advantage of attracting new leadership. The analogy to the establishment of the FAA should be looked at carefully because at that time the air was the new domain.

While each of the above alternatives has advantages and disadvantages, the committee believes the optimal approach would be to find a body of leaders who can cut across bureaucratic lines. The 1,000-ship Navy concept espoused by the CNO is based on many of the ideas required for a successful NSMS. The 1,000-ship Navy is the core from which that strategy will grow. However, the Navy with its vast network of international contacts and linkages still falls short in its ability to mount the necessary effort. A novel and extraordinary approach is needed to break through the international barriers abroad and interagency barriers at home. The deficiencies that exist across U.S. government entities mean that no single individual short of the President could coordinate and, especially, command across all the agencies involved, and he or she could not keep up with the effort nor could one or two subordinates. The committee's inclination is to urge that there be an independent, third-party study of the maritime problem focusing on security and probably on other aspects of the maritime domain, including the alternatives just outlined above. Hence, the CNO should exploit his access to the Commander in Chief by asking him to appoint a body of leaders with authority to find a solution that accords with his NSMS.

A Presidential Commission

Previously, when significant changes in government structure were needed, a Presidential Commission with appropriate terms of reference was set up.[36] Such a commission must draw heavily on the expertise found in several places. Govern-

[36]The historical precedent is the 1955-1957 Presidential Committee, which recommended the establishment of the Federal Aviation Administration (FAA) so as to "consolidate all the essential management functions necessary to support the common needs of the military and civil aviation of the United States." The rationale for the FAA was similar organizationally and functionally to the current need to remedy the shortfalls in roles, responsibilities, and authority for maritime security. Back then, after some years of start-up problems and trial and error, the consolidation of organizational and functional responsibility in a single agency worked very well for the nation and indeed for the world.

ment, industry, trade organizations, labor unions, and academia all are potential contributors. There needs to be international representation or, at a minimum, consultation with other countries, especially in view of the anticipated interactions with the Law of the Sea.

Coordination at the Strategic Level

There is a critical need for coordination at the strategic level for all programs and initiatives that come under the umbrella of U.S. programs supporting the partnerships and maritime security. Neither the Navy nor the USCG can operate effectively within the MSP without support from other U.S. agencies and departments that provide the policy framework from which to execute assignments across the spectrum of maritime security (see Figure 4.2). There is a need to identify and establish various levels of support and coordination for individual countries as well as entire regions based on different levels of partnership and needs.

Operational and Tactical Support

To effectively translate strategic decisions in Washington it will be necessary to designate the level of support to a specific country or region. It will require the coordination of domestic partners and resources. At the local level the U.S. ambassador and his country team would coordinate all the U.S. programs related to maritime security and safety in that country. If either the Navy or the USCG deploys units to work on maritime security missions in a specific region, those units must coordinate their activities with those of the country teams to ensure that they are supporting the local requirements as well as the COCOM's Theater Engagement Plan.

FINDING AND RECOMMENDATION

The trend during the past two decades toward globalization in the exploitation of natural resources and in the manufacturing sector has meant an increasing need for maritime transport. This need in turn results in growing coastal trade, transoceanic commerce, shipbuilding, port expansion, fuel consumption, and competition for offshore resources—including fish stocks—all of which have significant impact on national and international governance related to maritime safety, control, and security. The governance burden, especially as regards security, is already straining U.S. resources for protecting the country's own waters and ports. It is time to act on this understanding and prepare the nation and its prospective partners to deal with the growing task of maritime governance.

Establishing a regime such as that implied for MSP is an extensive and exceedingly complex task that needs to involve departments and agencies across the U.S. government. It needs to engage other participants in ways that transcend formal military and political alliances, and it needs to be seen by other countries not as a U.S. military initiative but as a way of fostering law and order at sea and thus the security of all participants. It is not clear that the existing MSPCC—despite some positive steps at the policy level—has adequate authority or mechanisms to fully realize MSP objectives as part of the national strategy. The situation bears a strong resemblance to the situation that faced the nation with respect to air transportation before the establishment of the FAA and the International Civil Aviation Organization.

Finding: The Chief of Naval Operations' initial 1,000-ship Navy concept has become a much larger concept of maritime security partnerships, attracting much international recognition and interest. It has grown beyond a U.S. Navy initiative into a critical matter for all agencies of the U.S. government that deal with international maritime relationships and trade.

Recommendation 12: The Chief of Naval Operations should recommend the appointment of an independent third party such as a presidential commission on maritime security governance tasked to recommend ways of strengthening the nation's maritime security policy, to define the roles and responsibilities of various U.S. government agencies and departments to better implement maritime security partnerships both domestically and internationally, and to move forward as suggested in the 11 other recommendations of this report.

Appendixes

A

Committee and Staff Biographies

Robert B. Pirie Jr., *Co-chair,* is an independent consultant with more than 40 years of expertise in Department of Defense (DOD) planning, programming, and budgeting. He served 20 years as a naval officer, culminating his service with 3 years in command of a nuclear attack submarine. He also served as assistant secretary of defense in the Carter administration, assistant secretary of the Navy and under secretary of the Navy in the Clinton administration, and acting secretary of the Navy from January until June of 2001. Mr. Pirie has also held a number of senior positions in the private sector, including that of president at Essex Corporation and vice president at the Institute for Defense Analyses.

David A. Whelan (NAE), *Co-chair,* is vice president and deputy general manager of Advanced Systems and chief scientist for Integrated Defense Systems at the Boeing Company's Phantom Works. His areas of expertise include defense research, development, and enabling technologies, such as autonomous vehicles and space-based, moving-target-indicator radar systems. Prior to joining Boeing, he served as director of the Tactical Technology Office at the Defense Advanced Research Projects Agency (DARPA). Dr. Whelan formerly held several positions of increasing responsibility with Hughes Aircraft. His high-technology development experience also included roles as a research physicist for the Lawrence Livermore National Laboratory and as one of four lead engineers assigned for the design and development of the B-2 Stealth Bomber Program at Northrop Grumman. He has served on numerous scientific boards and advisory committees, such as the Defense Science Board and the Air Force Scientific Advisory Board, and is a member of the Naval Studies Board of the National Research Council (NRC).

Noel K. Cunningham is currently CEO of the MARSEC Group, a maritime security consulting services firm. Mr. Cunningham is the retired director of operations for the Port of Los Angeles. In that capacity, he managed the port police department, port pilot services, homeland security, and emergency management divisions. Mr. Cunningham's background includes a career in law enforcement as a command officer in the Los Angeles Police Department; extensive experience in maritime and homeland defense and risk assessment; and experience with federal, state, and local laws applicable to cargo protection, pollution, vessel traffic control, and drug interdiction. He was formerly the chief of police for the Port of Los Angeles and was a member of the NRC Committee on the Role of Naval Forces in the Global War on Terror.

Henry H. Gaffney is director of the Strategy and Concepts Group in the Center for Strategic Studies at the Center for Naval Analyses Corporation (CNAC). His research interests range from military force structure to globalization; most recently, he examined military transformation, the changing nature of warfare through 2020, energy security, and global climate change. Prior to joining CNAC, Dr. Gaffney served for 28 years in the Office of the Secretary of Defense, where his activities focused on NATO and the Near East in security assistance affairs.

Gunther Handl is Eberhard Deutsch Professor of Public International Law at Tulane University. His expertise includes law of the sea, comparative law, international environmental law, transnational litigation, and the intersection of law, science, and technology. Professor Handl has served as a consultant to various international organizations and governmental agencies and as a special advisor in the legal advisor's office at the Austrian Ministry of Foreign Affairs. He also served as a professor of law at Wayne State University and as an associate professor of law at the University of Tulsa.

Thom J. Hodgson (NAE) is distinguished university professor in the Industrial and Systems Engineering Department at North Carolina State University. His expertise includes scheduling and logistics as well as modeling and optimization approaches, classic job shop and industrial scheduling, supply chain management, and military logistics. He has served on numerous scientific boards and advisory committees, including as chair of the NRC Committee on Evaluation of Manufacturing Vision and Strategies for the Production of the Crusader Artillery System.

James D. Hull retired from the U.S. Coast Guard with the rank of vice admiral and currently serves as a principal advisor on homeland security for the Security Strategies and Operations Group at Anteon Corporation. His background includes Coast Guard and interagency operations and capabilities, as well as maritime security and intercept operations. During his Coast Guard career, Admiral Hull served as commander of the Coast Guard's Atlantic area and the U.S. Maritime Defense Zone Atlantic. He has served on numerous scientific boards and advisory committees and was a member of the NRC Committee on the Role of Naval Forces in the Global War on Terror.

Harry W. Jenkins Jr. retired from the U.S. Marine Corps with the rank of major general and is currently an independent consultant. General Jenkins's background includes naval operations, mine countermeasures, and Marine Corps intelligence operations, in particular, its mission use of C4ISR systems. He formerly served as director of business development and congressional liaison at ITT Industries-Defense, where he was responsible for activities in support of tactical communications systems and airborne electronic warfare between the Navy, the Marine Corps, the Coast Guard, and the National Guard. During Operation Desert Storm, General Jenkins served as commanding general of the Fourth Marine Expeditionary Brigade. He is a member of the board of governors of the Marine Corps Association and a member of the Naval Studies Board.

Catherine M. Kelleher is a professor of public policy at the University of Maryland and a senior faculty associate at Brown University's Watson Institute. Her research interests include cooperative European defense and security policies, NATO relations, and international security and arms control. Dr. Kelleher served in the Clinton administration as the personal representative of the secretary of defense in Europe and as deputy assistant secretary of defense for Russia, Ukraine, and Eurasia. She has served on numerous scientific boards and advisory committees, including as vice chair, co-vice chair, and member of the Committee on International Security and Arms Control.

Jerry A. Krill is assistant director of programs at the Johns Hopkins University Applied Physics Laboratory (JHU/APL), where he oversees more than 400 programs and is also the laboratory's chief quality officer. His expertise includes weapons systems engineering, sensor and weapons networks, precision engagement and information-centric operations, missile defense, over-the-horizon missile command-and-control systems, and microwave technology. Previously, he served as head of the Power Projection Systems Department, program manager for the Air and Missile Defense Area, and supervisor of the Weapon Systems Engineering Branch. He has served on numerous scientific boards and advisory committees, including as a member of the NRC Committee on C4ISR for Future Naval Strike Groups.

Thomas V. McNamara is senior vice president of the Advanced Solutions Center at Textron Systems. His expertise includes intelligent autonomous systems, precision weapons delivery command and control, microelectromechanical systems development, guided munitions and missile technologies, Global Positioning

System antijam and ground control, and systems development and integration efforts for naval submersible and aircraft platforms. He served as a member of the NRC Committee on Distributed Remote Sensing for Naval Undersea Warfare; he is also a member of the Naval Studies Board.

Heidi C. Perry is division leader of mission systems at the Charles Stark Draper Laboratory. Her expertise includes guidance, navigation, and control; Global Positioning System antijam and ground control; precision weapons delivery command and control; guided munitions and missile technologies. Previously, Ms. Perry served as software engineering division leader and principal member of the technical staff at Draper.

Gene H. Porter is an independent consultant. His areas of expertise include national security planning and weapons systems development and defining the defense planning scenarios that are intended to guide the development of the U.S. military force structure. Mr. Porter formerly served as the director of acquisition policy and program integration at the Office of the Under Secretary of Defense for Acquisition. He has served on numerous scientific boards and advisory committees, including as chair of the NRC Committee for Mine Warfare Assessment. Mr. Porter is a member of the Naval Studies Board.

John S. Quilty is retired senior vice president and director of the C3I DOD Federally Funded Research and Development Center at the MITRE Corporation. His background includes supporting the technical requirements of the Army, Navy, Defense Information Systems Agency, Office of the Secretary of Defense, Office of the Joint Chiefs of Staff, and other members of the national security community. Mr. Quilty's recent work focused on support of DOD initiatives and activities to achieve improved C3I support to joint operations. He has served on numerous scientific boards and advisory committees, such as the Defense Science Board. Mr. Quilty formerly served as a member of the NRC Committee on the Role of Naval Forces in the Global War on Terror and is a member of the Naval Studies Board.

J. Paul Reason retired from the U.S. Navy with the rank of admiral after 34 years of service and is currently an independent consultant. His background includes naval and joint operations, as well as DOD planning, programming, and operations. In his last position, he served as commander in chief, U.S. Atlantic Fleet, where his responsibilities included the training, maintenance, and readiness of naval forces deployed to the Mediterranean and Caribbean seas, South America, and the Persian Gulf. He was also responsible for the operations of most U.S. Navy bases and facilities along the East and Gulf coasts of the United States, and in Puerto Rico, Cuba, and Iceland. ADM Reason is a member of the Naval Studies Board.

Nils R. Sandell Jr. is vice president and general manager of BAE Systems Advanced Information Technologies. His expertise includes automatic target recognition; sensor fusion; sensor resource management; battle management; and command, control, and communications. He formerly served as an associate professor at the Massachusetts Institute of Technology, where he lectured in the areas of estimation and control theory, stochastic processes, and computer systems. Dr. Sandell has served on numerous scientific boards and advisory committees, including as co-chair of the NRC Committee on C4ISR for Future Naval Strike Groups. He is a member of the Naval Studies Board.

H. Eugene Stanley (NAS) is university professor, professor of physics, and director of the Center for Polymer Studies at Boston University. His expertise includes sensors and polymeric materials; theory of phase transitions and critical phenomena for a wide range of systems, including polymers; and applications of statistical mechanics to biology, economics, and medicine. Dr. Stanley was a member of the NRC Panel on Nonlinear Science.

John P. Stenbit (NAE) is an independent consultant whose expertise includes system architectures for complex military and communication systems and systems engineering of information systems. Mr. Stenbit formerly served as assistant secretary of defense for networks and information integration and DOD chief information officer. Prior to serving in the DOD, he served as executive vice president at TRW, Incorporated. Mr. Stenbit has served on numerous scientific boards and advisory committees, including as a member of the NRC Committee on C4ISR for Future Naval Strike Groups. He is also a member of the Naval Studies Board.

Elihu Zimet is a distinguished research professor in the Center for Technology and National Security Policy at the National Defense University (NDU). His background includes naval science and technology, including kinetic and nonkinetic effects, and low-observable and counter-low-observable technologies. Prior to joining NDU, he served as head of the expeditionary warfare science and technology department at the Office of Naval Research. Dr. Zimet served on the NRC Committee on the Role of Naval Forces in the Global War on Terror; he is also a member of the Naval Studies Board.

Staff

Charles F. Draper is director of the NRC's Naval Studies Board. Before joining the NRC in 1997, Dr. Draper was the lead mechanical engineer at S.T. Research Corporation, where he provided technical and program management support for satellite Earth stations and small satellite design. He received his Ph.D. in mechanical engineering from Vanderbilt University in 1995; his doctoral research was conducted at the Naval Research Laboratory (NRL), where he used an atomic-force microscope to measure the nanomechanical properties of thin-film materials. In parallel with his graduate student duties, Dr. Draper was a mechanical engineer with Geo-Centers, Inc., working on-site at NRL on the development of an underwater X-ray backscattering tomography system used for the nondestructive evaluation of U.S. Navy sonar domes on surface ships.

Arul Mozhi is senior program officer at the NRC's Division of Engineering and Physical Sciences. Prior to joining the NRC in 1999, Dr. Mozhi was senior scientist and program manager at UTRON, Inc., a high-tech company in the Washington, D.C., area, working on pulsed electrical and chemical energy technologies applied to materials processing. From 1989 to 1996, Dr. Mozhi was a senior engineer and task leader at Roy F. Weston, Inc., a leading environmental consulting company, working on long-term nuclear materials behavior and systems engineering related to nuclear waste transport, storage, and disposal in support of the U.S. Department of Energy. Before 1989 he was a materials scientist at Marko Materials, Inc., a high-tech firm in the Boston area, working on rapidly solidified materials. He received his M.S. and Ph.D. degrees (the latter in 1986) in materials engineering from the Ohio State University and then served as a postdoctoral research associate there. He received his B.S. in metallurgical engineering from the Indian Institute of Technology in 1982.

B

Sea Lanes of Commerce in the Various Regions of the World

Globalization in the 21st century has forced into keen focus the absolute imperative for an ability to assure free and peaceful access to the sea. The U.S. economy—in fact, all economies of all developed and developing nations and multinational corporations—are more reliant than ever before on global trade for their prosperity. The exchange of raw materials, product components, and finished goods by sea conveyance has paralleled the expanding global economy. But this exchange requires free and uninterrupted use of the seas, which has seen a largely peaceful environment for the past 50 years due in large part to the maritime dominance of the United States and its allies and friends.

Maritime security partnerships (MSP) may become the means by which all nations contribute to maintaining the freedom of the seas at the same time as they protect their homelands.[1]

Without assured freedom of the seas, global trade and global economies could be hindered. Consequently, all users of the sea for commerce should embrace and support such initiatives that will protect the seas from criminal activity and disruption.

In stable regions of the world, where maritime trade is mature and follows established routes, commodities, and even schedules, evolving technologies have been applied to optimize the generation of data that immediately highlight any disruption to normal commerce.

Multinational corporations, shipping lines, coast guards, port authorities, and any number of government entities should find it in their interest to invest

[1]ADM J. Paul Reason, Commander in Chief, U.S. Atlantic Fleet, with David G. Freymann. 1998. *Sailing New Seas,* The Newport Papers, Thirteenth in the Series, Naval War College, Newport, R.I.

their resources to curb illegal and disruptive activity, ensuring the free conduct of maritime trade. However, in particular areas of the world, usually coastal, illicit maritime trade may be violating the 2005 Protocols to the 1988 Suppression of Unlawful Acts (SUA) Convention and its Protocol.[2]

The list of shared, complex challenges is long; such challenges usually grow out of conditions whereby regions featuring stable governments, rising standards of living, increased trade, and network connectivity are pulling away from regions of the world where nations are plagued by politically repressive regimes, weak economies, widespread poverty, disease, and a lack of adequate medical care. The challenges include but are not limited to terrorism, weapons proliferation, trade disruption, piracy, the drug trade, human smuggling, illegal immigration, and organized crime. They could also include environmental attacks, illegal fishing, competition for natural resources such as oil for developing countries or, in some areas, the growing shortage of water, which can lead to famine. The natural disasters that occur regularly around the world are another challenge to which maritime forces have to react because only they may be able to reach hard-hit areas to deliver assistance. The most recent example of this was the tsunami that devastated large parts of Indonesia, Thailand, and the island chains in the Indian Ocean and Andaman Sea. Just as graphic was the maritime support in response to Hurricane Katrina on the Gulf Coast of the United States.

The shared complex challenges chart (Figure B.1) is a 30-day picture of the ocean-going traffic that moves on the high seas as well as in the sea lanes of commerce (SLOCs) and choke points around the world. Each dot on that figure represents one ship, and the trade routes are very clear from the density of shipping around the major continents. Many of the choke points through which much of the trade flows are in regions where countries are in various states of development or have tenuous relations with each other or with the more developed countries.

Open SLOCs are critical for world trade and for global maritime security. What follows is a general description of the key SLOCs by region.

THE PACIFIC AND SOUTHEAST ASIA

SLOCs in the Indonesian archipelago and the South China Sea remain critical choke points in that almost half of the world's shipping passes through these waters. They include the Strait of Malacca, the Strait of Sunda, and the Strait of Lombok. All three provide entrances from the south to the South China Sea. The Strait of Makassar could also be considered a choke point. Each is important to the world trade system (see Figure B.1).

- From an economic and strategic prospective, the Strait of Malacca is one

[2]Protocol 2005 to the Convention for the Suppression of Unlawful Acts against the Safety of Maritime Navigation, and Protocol of 2005 to the Protocol for the Suppression of Unlawful Acts against the Safety of Fixed Platforms Located on the Continental Shelf.

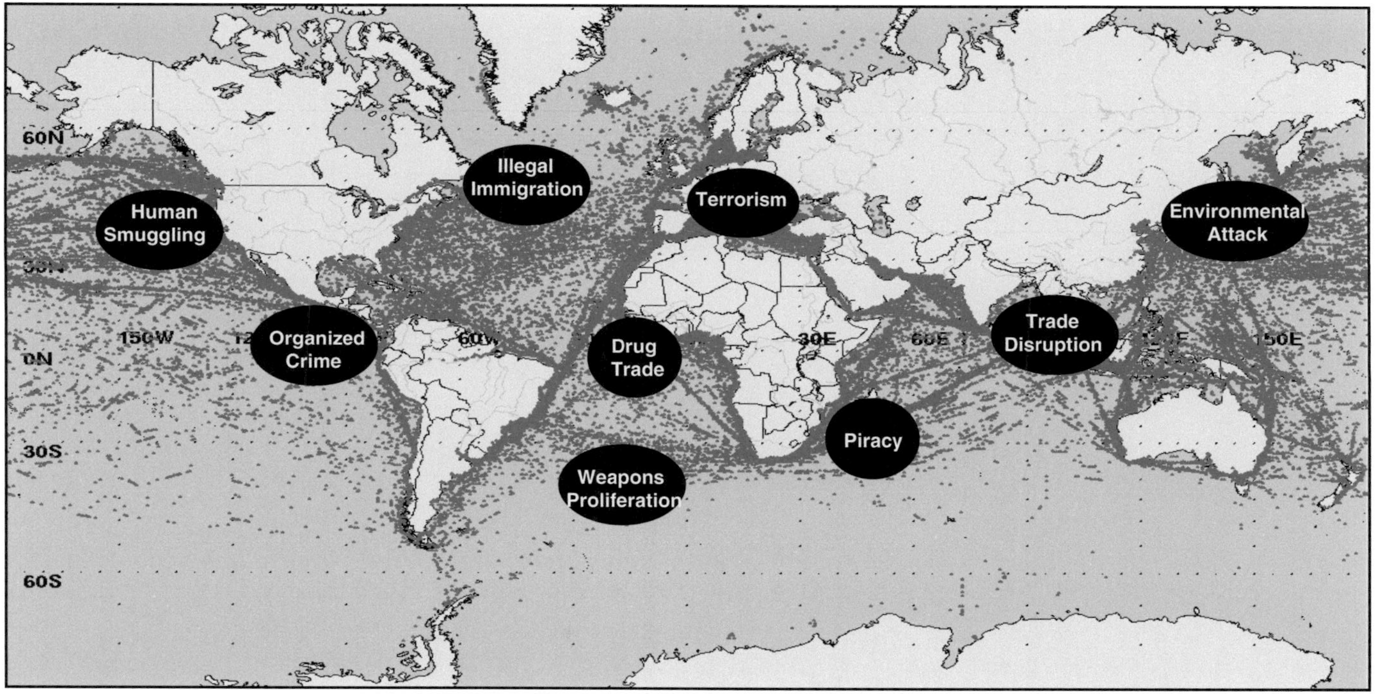

FIGURE B.1 Complex shared challenges. SOURCE: ADM Henry G. Ulrich III, USN, Commander, U.S. Naval Forces Europe, Commander, Allied Joint Forces Command, Italy, "Complex Shared Challenges," presentation to the committee, Naples, Italy, March 29, 2007.

of the most important SLOCs in the world. Over 50,000 vessels of all sizes pass through here annually between the Indian Ocean and the Pacific. Over a quarter of all oil carried by sea moves through this strait. A terrorist attack or increased piracy in this waterway could have a large-scale economic impact on the region as well as the world. Any successful attack or blockage could dramatically raise insurance rates for ships transiting this area or could force ships to detour well out of their way, causing major shortages of crude oil or dry bulk cargoes like iron ore or coal. The result could be higher freight rates as well as a disruption of world markets.

- The Sunda Strait passes between the Indonesian islands of Java and Sumatra. It connects the Java Sea with the Indian Ocean. Sunda could be used as an alternative if the Strait of Malacca were closed for some reason; however, its narrowness at points as well as oil rigs off the Java coast could make it unsuitable for large commercial vessels.
- The Lombak Strait connects the Java Sea to the Indian Ocean and is located between the islands of Bali and Lombak in Indonesia. The Makassar Strait runs between the islands of Borneo and Sulawesi. Much of Australia's export trade carried in ships to northeast Asia goes through both of these two deepwater passages in Indonesia. Depending on a ship's destination, it branches off toward the Philippine Sea or into the South China Sea. Many other countries use these waterways as well.
- Many nations in Southeast Asia are islands or have extended coastlines. Their land transport infrastructure is not well developed, although seaborne imports are growing and thereby increasing the use of SLOCs for interisland trade. A key point here is that the myriad of islands in this region make it extremely difficult to provide adequate coverage for situational awareness in support of maritime security requirements. The current exception is the "electronic highway" in the Malacca Strait.

THE MIDDLE EAST, THE INDIAN OCEAN, AND AFRICA

This large expanse of ocean features some of most critical SLOCs and choke points in the world. Shipping traffic is generally secure in the open ocean but not in some of the coastal areas. Several choke points are bordered by states ruled by regimes that are weak politically, corrupt, and more or less hostile to the United States and other nations. Many are threatened by terrorism, piracy, the drug trade, or smuggling (see Figure B.1).

- The Strait of Hormuz is a strategically important, very narrow channel between the Persian Gulf, the Gulf of Oman, and the Indian Ocean. It is a vital shipping lane for petroleum tankers traveling to or from the Far East, Africa, or Europe. Over 25 percent of the world's oil supply passes through this strait, which is bordered by Iran to the east and Oman and the United Arab Emirates to the west and south. Transit problems in this waterway would lead to widespread trade disruption in the oil markets as well as potentially severe economic consequences in the countries that depend on oil from that region.

- Bab El-Mandeb at the entrance to the Red Sea, a vital SLOC and choke point, is bordered by Yemen and Saudi Arabia to the east; Djibouti, Ethiopia, and the Sudan to the west; and Somalia to the south. The countries in this area are all emerging and are havens for terrorism, weapons smuggling, the drug trade, and piracy. All commercial shipping, including petroleum tankers, moving from the Indian Ocean to the Mediterranean and Europe must pass through the Bab El-Mandeb and the Red Sea to the Suez Canal. Likewise, most shipping coming from Europe to the Middle East, India, and the Pacific must pass through this waterway. Disruption at any point along this SLOC could force shipping companies to move their cargoes by sea from Europe to Asia around South Africa, causing major trade disruptions, economic chaos, and drastic increases in insurance premiums for the shippers.
- The Mozambique Channel lies between Madagascar to the east and Mozambique to the west, along the East African coast. It is an important shipping route for countries bordering the Indian Ocean to and around the southern tip of Africa and into the South Atlantic. The channel is wide and deep and consists of island groups that are considered strategically important from the standpoint of maritime security. Piracy, the drug trade, illegal fishing, human smuggling, and terrorism are all problems in this area. All of the countries in this region are fragile and have weak economies and poor maritime security.
- The area along the western coast of Africa, known as the Gulf of Guinea, is rich in natural resources (sweet crude oil) and bordered by several countries with weak governments, corruption, struggling economies, and large ungoverned areas. It is plagued by militant violence, illegal fishing, piracy, and poverty. The Gulf of Guinea is currently the third largest source of oil imports to the United States, and it is projected to be one of the world's top four oil producers by 2020. The legal framework for maritime law is inadequate or nonexistent, and few of the countries bordering the Gulf have the capacity to provide maritime security. Ships moving along the west coast of Africa pass through the Gulf of Guinea.

EUROPE AND THE BLACK SEA

This region is characterized by stable countries with viable economies that stretch from the Mediterranean to the Baltic and the Scandinavian countries. To the east of the Mediterranean and the Black Sea are some stable countries as well as others in varying stages of development. Trade in the SLOCs of the Mediterranean, the English Channel, and the Baltic is relatively normal; however, the drug trade, the potential for terrorism, illegal immigration, and smuggling are concerns for maritime security. Choke points in this region could be the Bosporus, the Strait of Gibraltar, and the Skagerrak and Kattegat, which lie between the North Sea and the Baltic. Of the three, the Bosporus, which sits between the Mediterranean and the Black Sea, may be the most critical as it is the only water route from the Black Sea countries to the Sea of Marmora, the Dardanelles, and the Aegean Sea (see Figure B.1). Much of the oil coming from the Caspian Sea region passes through this waterway. The Bosporus is the world's narrowest strait that is used for international navigation, and there have been many conflicts over it in modern times. However, there are international treaties in place that govern the use of the waterway. By treaty, Turkey controls the Bosporus as well as the Dardanelles to the west.

THE WESTERN HEMISPHERE AND THE CARIBBEAN

The SLOCs between the Pacific and Atlantic and both North and South America are open, and commercial shipping moves freely with a minimum of interference. Maritime security concerns center on the threats of terrorism, smuggling, weapons proliferation, and the drug trade. In the Caribbean the primary concerns are the drug trade and illegal immigration. The countries of Latin America are mainly stable, and the maritime commerce there depends heavily on the Panama Canal, the quickest route from the Atlantic to the Pacific for world trade.

The Panama Canal crosses the Isthmus of Panama in Central America. It is a key conduit for international shipping: More than 14,000 ships pass through it annually over the 70-mile route. If the Panama Canal were ever closed to commercial shippers, it would mean a long voyage around South America that would disrupt trade, slow down economies, and drive up insurance premiums for the shipping companies (see Figure B.1).

While the Panama Canal has enjoyed considerable success, there could be problems in the future. The volume of imports from Asia now moving through the canal for ports on the East Coast of the United States and other ports in the hemisphere is increasing. The number of transits is down; however, the total tonnage capacity has gone up, from 227 million tons in FY1996 to almost 296 million tons in 2006. Canal authorities have widened and modernized portions of the waterway, which has increased efficiency, but it is expected to soon reach its maximum capacity. With larger and larger ships lining up in the assembly areas at both ends of the canal, the potential for accidents and terrorism goes up. Destruction of one of the critical locks along the route by terrorists would present a major challenge for the authorities responsible for security in and around this waterway.

* * * * * * * *

SLOCs in all regions of the world are critical for the free movement of commercial shipping and for world trade. Today these lanes are not threatened by wars between countries, but the potential for disruption of commerce is nonetheless significant if maritime security frameworks are not in place in certain regions of the world. Individuals or groups that want to disrupt trade along any of these routes could do so at many points in the maritime domain. Covering all of the critical SLOCs as well as the choke points with adequate surveillance systems having links to maritime security forces is an onerous responsibility and very expensive. The question becomes who will pay for the costs of such systems and who will create and coordinate the policies that will be the legal foundation for the surveillance plans. Achieving maritime domain awareness (MDA)[3] has to be the first step following regional partnerships and collaboration.

[3]Department of Homeland Security, 2005, *The National Plan to Achieve Maritime Awareness for the National Strategy for Maritime Security* (Washington, D.C.), October, p. 1, defines MDA as "the effective understanding of anything associated with the maritime domain that could impact the security, safety, economy or the environment of the United States."

C

The International Legal Framework

Chapter 2 briefly discussed the need for an international legal framework for maritime security partnerships (MSP). This appendix provides more details and the committee's observations on such a framework.

THE STRUCTURE OF MARITIME GOVERNANCE: RESTRAINTS OR EMPOWERMENT?

A stark reminder of the crucial significance of the legal parameters applicable to MSP is provided by a United Nations report that notes, succinctly, "any measures taken to prevent terrorist acts against shipping, offshore installations and other maritime interests must be in conformity with international law, including UNCLOS."[1] Considering the importance that states in general ascribe to the United Nations Convention on the Law of the Sea (UNCLOS) as having a "universal and unified character" whose "integrity needs to be maintained,"[2] it follows that activities affecting the oceans not only will have to pass muster in accordance with relevant substantive rules and standards of UNCLOS but also will need to comply with the general understanding among states of how UNCLOS should be adapted, if it becomes necessary. This understanding suggests a process that is

[1]United Nations, *Report of the Secretary-General on Oceans and the Law of the Sea,* New York, A/62/66, Paragraph 81, March 12, 2007, p. 28. Thus, it is generally agreed that UNCLOS "sets out the legal framework within which all activities in the oceans and the seas must be carried out and is of strategic importance as the basis for national, regional and global action and cooperation in the marine sector. . . . " United Nations, Oceans and the Law of the Sea, New York, Resolution No. A/RES/61/222, Preamble, March 16, 2007, p. 1.

[2]See, for example, United Nations Resolution No. A/RES/61/222, Preamble, March 16, 2007.

centered, first and foremost, on the states that are themselves parties to UNCLOS or, exceptionally, on an equally broad-based multilateral approach that is inclusive as well as transparent.

UNCLOS itself does not comprehensively or even specifically address the matter of maritime security. Instead, it features a number of limited rules that speak directly to issues of maritime security, such as those in Arts. 101-105 (on piracy), Art. 110 (on boarding of foreign flag vessels on the high seas without the consent of the flag states), and Art. 111 (on hot pursuit). More importantly, however, in restating the all-important customary international legal principles of jurisdiction over ocean spaces, it determines the allocation of rights and obligations between flag states on the one hand and coastal and port states on the other regarding security-related activities in these maritime zones.

A second tier of relevant normative standards, either in place today or about to emerge, specifically addresses security-related activities on the oceans. It includes key post-9/11 international legal developments: the various amendments to the International Convention for the Safety of Life at Sea (SOLAS), including, in particular, Chapter XI-2 on "special measures to enhance maritime security"; the adoption of the International Ship and Port Facility Security (ISPS) Code; UN Security Council resolution 1540; and some international agreements that have yet to enter into force, namely, the 2005 Protocols to the 1988 SUA Convention and its Protocol[3] and the International Convention on the Suppression of Acts of Nuclear Terrorism.[4] Other global international legal instruments—the 1988 UN (Vienna) Convention against Illicit Traffic in Narcotic Drugs and Psychotropic Substances; the 2000 UN Convention against Transnational Organized Crime;[5] the Protocol against the Smuggling of Migrants by Land, Sea and Air supplementing the UN Convention against Transnational Organized Crime;[6] and the UN Convention against Corruption[7]—represent additional building blocks of a maritime-security-specific global legal architecture, yet to be completed.

The aggregate effect of these legal agreements and instruments, whether in force or not, is to lend political support to the concept of MSP. As the embodiment of specific international legal authority bearing on MSP, they cover some of the activities that directly promote maritime domain awareness (MDA) and associated responses. At the same time, however, they also signal clearly the limits of states' authority to take action on their own or on a limited regional basis in order to maximize MDA and related response options. Regional-security-related agree-

---

[3]Protocol of 2005 to the Convention for the Suppression of Unlawful Acts against the Safety of Maritime Navigation, and Protocol of 2005 to the Protocol for the Suppression of Unlawful Acts against the Safety of Fixed Platforms Located on the Continental Shelf.

[4]United Nations, International Convention for the Suppression of Acts of Nuclear Terrorism, New York, Resolution No. A/RES/59/290, Annex, April 15, 2005, pp. 1-13.

[5]United Nations, Resolution No. A/RES/55/25, Annex I, January 8, 2001.

[6]United Nations, Resolution No. A/RES/55/25, Annex III, January 8, 2001.

[7]United Nations, Resolution No. A/58/422, October 7, 2003.

ments, such as the recently concluded ASEAN Convention on Counterterrorism,[8] play a similar dualistic role: While aiming at fostering cooperation on international security by participating states, they do not seek to create new international legal authority for security measures among the states concerned.[9] Instead, the agreements tend to remain faithful to the traditional multilateral/global allocation of rights and obligations of states. They thereby confirm indirectly the existing global governance structure, pursuant to which international legal change of a general nature requires the general participation of states, if not their general consensus on the outcome.

Finally, bilateral arrangements, such as the counterdrug agreements the United States has entered into with, for example, Caribbean and Latin American nations,[10] admittedly often do change—bilaterally, or *inter partes*—the general rules of law that might apply to maritime security operations. However, these agreements cannot be considered in isolation from the multilateral legal platforms on which they are based and that provide the specific enabling authority or political coverage for individual states to enter into these bilateral agreements in the first place.[11] Thus, the "bilaterals" do not in and of themselves provide a legal basis for expanded general maritime security cooperation among the states concerned, nor do they necessarily represent a model that could be readily emulated elsewhere in the world. By the same token, various informal understandings and non-law-based practical cooperative arrangements, such as the Proliferation Security Initiative (PSI),[12] are undoubtedly useful in facilitating and promoting maritime security cooperation in general. However, they have not created new international legal authority where previously none existed.[13] These kinds of arrangements, therefore, should not be mistaken for representing suitable substitutes for the type of explicit international legal authority, multilaterally agreed upon, that some MSP-related activities unquestionably require.

---

[8]Done at Cebu, Philippines, January 13, 2007.

[9]Note, e.g., Articles III-V of the Convention.

[10]See infra note 87.

[11]For example, the cooperative counterdrug agreements and arrangements that JIATF-S relies on are in turn embedded in the 1988 UN (Vienna) Convention against Illicit Traffic in Narcotic Drugs and Psychotropic Substances and the 2000 UN Convention against Organized Crime. Similarly, the Cooperating Nations Information Exchange System (CNIES) also derives support from the Inter-American Convention on Mutual Assistance in Criminal Matters, the UN Convention against Corruption, and so on.

[12]Or, the Global Initiative to Combat Nuclear Terrorism: See Statement of Principles by Participants in the Global Initiative to Combat Nuclear Terrorism, at <http://www.state.gov/r/pa/prs/ps/2006/75405.htm>. Accessed on April 17, 2008.

[13]Indeed, as a rule, they are not intended to change applicable international legal rules, nor have they incidentally brought about such changes. See, for example, the preambular sentence of Principle 4 of PSI Statement of Interdiction Principles, which expressly requires that interdiction efforts be consistent with participating states' "obligations under international law and frameworks."

Specific Observations on Structure-of-Maritime-Governance Issues

To date there exists no overarching, comprehensive legal basis for maritime-security-specific measures. International law continues to provide only a partial or fragmented basis for the specific measures that might be indispensable for launching a system of effective MSP. This situation represents both an opportunity and a handicap. It is an opportunity because it permits participants to engage in incremental steps toward implementing MSP with all the attendant political benefits of a "small steps, deliberate speed" approach. At the same time, it is a handicap because specific changes in the applicable rules of international law, whenever necessary, cannot readily be justified as mere "measures of implementation" by reference to a preexisting generic legal framework document or instrument on maritime security. Instead, proposed changes will have to be vetted individually as to their international legal acceptability, the existence of international authority, and, in particular, their compatibility with UNCLOS.

As the constitution of the oceans, UNCLOS provides the fundamental legal framework for MSP. Thus any MSP-related proposals that imply changes in the maritime rules of the game bring into play UNCLOS. However, as the chief proponent of the MSP initiative, the United States is not at present a party to UNCLOS.

The utility of bilateral legal understandings and formal agreements—as, for example, between the United States and Caribbean and Latin American nations—in fostering a web of single- or multiple-issue-focused security partnerships is self-evident. Operationally, bilateral or (limited) regional MSP is likely to be the most effective model of security cooperation. But, such arrangements will normally not be feasible unless they are founded on a solid multilateral legal basis.

The coalition-of-the-willing model à la PSI (which, for example, expressly disavows any intention to change traditional international law regarding the boarding of foreign flag vessels) does not obviate the need for recourse to proper multilateral processes and settings to effect legal change.

It should be clearly understood, finally, that the scope of existing, emerging, or even proposed maritime-security-related international legal measures tends to be limited by two factors. First, generally speaking, public vessels—that is, warships and other vessels owned or operated by the government of a country and that are not engaged in commercial service[14]—will be either exempt from MDA-enhancing international rules and regulations or, if such rules do apply, will not be subject to the enforcement jurisdiction of any state other than their own flag state. Second, this restricted focus on commercial vessels, barges, and so on is

[14]Under U.S. federal law, the distinction between commercial and noncommercial vessels tends to turn on the commercial vs. governmental nature of the activity the vessel is engaged in. See, for example, the definition in 33 CFR §160.24 and the by now traditional test for immunity under the Foreign Sovereign Immunities Act. By contrast, under UNCLOS itself the purpose of the activity will be the decisive criterion. See, for example, Arts. 31-32 in UNCLOS and the United Nations Convention on Jurisdictional Immunities of States and Their Property, Article 16.

exacerbated by the fact that most international legal measures target vessels of a minimum size only—usually above 300 GT. As a result, the very large number of vessels that could be of concern from a maritime security perspective will not be covered by relevant applicable international standards or regulations or by some of the important standards that are about to become operational.[15]

Since MSP would benefit from a generic, maritime-security-specific legal endorsement or declaration by states of basic principles and objectives and the designation of a lead international agency or organization, the Department of State (DoS) and the Department of Homeland Security (DHS)—of which the U.S. Coast Guard (USCG) is a part—need to support a broad-based diplomatic effort to this end. Such an effort might aim at the adoption of a resolution by the United Nations General Assembly or the Security Council.

While it is generally accepted today that most of the provisions of UNCLOS are part of customary international law, and as such binding upon and benefiting the United States, U.S. ratification of UNCLOS would be an important step in support of MSP and would give the United States a place at the table in UNCLOS-based decision-making bodies and related processes. Thus the committee concurs with the President, the Secretary of Defense, the Chairman of the Joint Chiefs of Staff, the Secretary of the Navy, the Chief of Naval Operations (CNO), the Commandant of the Marine Corps (CMC), and the Commandant of the Coast Guard (CCG) regarding ratification of UNCLOS at the earliest possible time.

The United States would benefit from addressing other states' concerns about the legitimacy of MSP by taking steps to bolster its multilateral and global credentials, by acknowledging the essentially multilateral nature of many of the tasks to be addressed, and by supporting the choice of multilateral and formal legal settings and forums, as appropriate, to ensure a transparent and inclusive process of review of the law of the sea and its adjustment if necessary.

Various international organizations already have extensive security-related portfolios that bear on MSP activities. This is true in particular of the International Maritime Organization (IMO), the United Nations Food and Agriculture Organization in relation to fisheries conservation and management (monitoring of fishing vessels), the World Customs Organization (WCO) in relation to container security, and Interpol. MSP-related efforts by U.S. stakeholders and agencies need to be tightly coordinated with the efforts of these organizations; indeed, the United States must secure their active involvement and draw on their relevant expertise and capacity.

All U.S. stakeholders in MSP would benefit by fully supporting the DHS (that is, the USCG) efforts to address the issue of security for small vessels by working with other states toward reducing, within the ambit of the IMO and FAO, the threshold for applicability of the relevant international maritime-security-related standards and regulations. Particularly important are efforts to lower the

[15]A case in point is the new Long-Range Identification and Tracking (LRIT) system.

threshold to below 300 GT (or appropriate units of length[16]) in any new maritime-security-related international legal instruments.

INTERNATIONAL LEGAL IMPLICATIONS OF THE IMPLEMENTATION OF KEY MDA-RELATED OBJECTIVES

The implementation or operationalization of specific MDA-related objectives essential to making MSP effective will raise a number of international legal issues. This is notably true for proposals to enhance cargo and container security; to raise the situational awareness of port and coastal states regarding vessels of interest by improving global vessel tracking capabilities; to expand port and coastal states' rights to access or right to vessel-related information; and to facilitate the sharing of maritime information, assisted by global vessel/port databases, and so on. It is equally true of measures to facilitate the boarding, including the nonconsensual boarding, of foreign flag vessels in areas or situations not subject to national jurisdiction and control as a means to acquire or verify relevant information about vessels of interest.

Cargo and Container Security—Off-shoring of Security Measures

With more than 11 million containers estimated to enter the United States annually, incoming ships and their foreign-origin cargo and containers pose a very significant security threat to the country. Recognizing this threat and that national security would best be served by addressing it at the point of loading in foreign ports, Customs and Border Protection (CBP), an arm of DHS—in direct response to the terrorist attacks of 9/11—launched the Container Security Initiative (CSI) in 2002. Its objective is to place U.S. customs agents at foreign ports for the purpose of identifiying and prescreening U.S.-bound high-risk containers before they are shipped to the United States. To date, the United States has entered into bilateral agreements to cooperate on customs and container security, with at least 50 foreign ports now involved in a CSI regime.[17]

CSI was initially conceived as, and thus far has been operated as, a program for the selective screening of U.S.-bound containers identified as a potential threat. However, full screening of all cargo destined for the United States through nonin-

[16]Note, for example, that some regulations of the European Community, such as those that apply to systems for monitoring fishing vessels, use a minimum (vessel) length metric.

[17]CSI is a reciprocal program, offering participating countries the opportunity to send their customs officers to major U.S. ports to inspect oceangoing, containerized cargo to be exported to their countries. Japan and Canada currently station their customs personnel in some U.S. ports as part of the CSI program. Likewise, CBP shares information on a bilateral basis with its CSI partners. The most recent addition to this growing list of CSI ports is Jawaharlal Nehru Port in India, which signed on to CSI in July 2007.

vasive inspection is now set to become the modus operandi[18] even though critics have raised legitimate doubts about the operational and/or economic feasibility of such an approach.[19] A parallel program, the Department of Energy's (DOE's) Megaport Initiative,[20] aims to deter, detect, and interdict trafficking in special nuclear materials and other radioactive materials by providing foreign commercial ports with U.S.-supplied and jointly operated technology and equipment.[21] Finally, an initiative complementary to CSI is the voluntary government-private sector program, the Customs-Trade Partnership Against Terrorism (C-TPAT).[22] It aims at securing the supply chain while expediting the cargo by establishing a system of trusted (or certified) agents in the international supply chain, such as importers, brokers, freight forwarders, and carriers, who would benefit from fast-tracking through stateside customs and security checks. A similar system based on the concept of an "authorized economic operator"[23] is scheduled to enter into force on a European-Community-wide basis next year.[24]

One of the by-products of CSI is the 24-hour cargo rule,[25] which requires sea carriers and nonvessel operating common carriers to provide CBP with a detailed description of the contents of a sea container bound for the United States 24 hours

---

[18]Following the recommendation of the 9/11 Commission, the Improving America's Security Act of 2007 will require the screening of all vessels and their cargo destined for the United States by 2014. Additionally, the SAFE Port Act, § 121(a), already imposes a requirement to scan for radiation—albeit in U.S. ports—all containers entering the country.

[19]See, for example, "Bill to Scan All Containers Entering the U.S. Will Cause Chaos, Say Importers," *Financial Times*, July 26, 2007, p. 1. Additionally, the European Community has attacked the bill as not cost-effective. See "Brussels Attacks American Plan to Scan Shipping Containers," *Financial Times*, August 3, 2007, p. 4.

[20]It is being administered by the U.S. National Nuclear Security Administration.

[21]A first such arrangement was entered into with the Bahamas in 2005. Since then several other countries have joined this initiative, including China, Jamaica, the Netherlands, Oman, and Singapore.

[22]See the Security and Accountability for Every Port Act of 2006, Public Law 109-374 [SAFE Port Act], § 211. The CBP's Customs–Trade Partnerships Against Terrorism (C-TPAT) program has a counterpart in the European Union's "authorized economic operator" program, with both programs aiming at eventual mutual recognition of nationally certified measures for security and facilitation of trade. Note also complementary industry efforts by companies like Siemens and General Electric to prevent tampering with shipping containers.

[23]The EC regulation implements the World Customs Organization's (WCO's) Framework of Standards to Secure and Facilitate Global Trade; § 2.3 (in footnote 1 on p. 6) defines "authorized economic operator" as "a party involved in the international movement of goods in whatever function that has been approved by or on behalf of the national Customs administration as complying with WCO or equivalent supply chain security standards."

[24]See Art. 14 of the Commission Regulation (EC) No. 1875/2006 of December 18, 2006, amending Regulation (EEC) No. 2454/93 laying down provisions for the implementation of Council Regulation (EEC) No. 2913/92 establishing the Community Customs Code.

[25]See 2005 CFR Title 19, Part 4—Vessels in Foreign and Domestic Trade, § 4.7 (2), according to which the incoming carrier must file on behalf of any vessel subject to a cargo declaration requirement as a condition for entry into a U.S. port the CBP-approved electronic equivalent of the vessel's cargo declaration 24 hours before the cargo is laden aboard the vessel at the foreign port.

before the container is loaded on board a vessel. The rule allows CBP officers to analyze the information on container contents and identify potential terrorist threats before the U.S.-bound container is loaded at the foreign seaport, not after it arrives in a U.S. port.

On a global, multilateral level, the December 2002 SOLAS Conference expressly recognized that maritime security was intrinsically tied to container security. Emphasizing that the intermodal and international nature of container movements necessitated ensuring the security of the entire supply chain, the Conference called on the WCO to address container security as a matter of urgency.[26] WCO, as the international organization with primary responsibility for supply chain security,[27] has since adopted the Framework of Standards to Secure and Facilitate Global Trade,[28] which promotes customs-to-customs and customs-to-business networks to improve the security of closed transport units.[29] Its core elements include the harmonization among participating states of advance electronic cargo information requirements and a commitment by countries joining the Framework to accede to requests by authorities of the destination state to perform an outbound inspection of high-risk containers and cargo.[30]

Specific Observations

Cargo and container security measures might be perceived as disproportionately benefiting the United States, and their propagation in the context of maritime security partnerships is therefore regarded as impolitic. But the very fact that the IMO, the WCO, other international organizations, and the European Community[31] have begun to address the issue demonstrates the wider international, indeed global, significance of container security: A serious security incident involving cargo or containers anywhere might have catastrophic consequences for maritime trade everywhere.

Cargo and container security is also apt to raise an issue of delimitation, namely, the question of which aspects of container security come properly within the ambit of MSP. Many security-sensitive cargo operations take place outside ports—that is, they involve shoreside segments of the supply chain, such as

[26]See Conference Resolution 9 on "Enhancement of Security Cooperation with the World Customs Organization (Closed Cargo Transport Units), Doc. SOLAS/CONF.5/34, Annex 2, 13.

[27]In 2007 a joint IMO Maritime Safety Committee/Facilitation Committee working group agreed that the WCO, rather than IMO, had primary responsibility for "supply chain security."

[28]WCO, Framework of Standards 6; see also WCO, Customs Guidelines on Integrated Supply Chain Management, June 2004.

[29]ILO/IMO Code of Practice on Security in Ports supplements WCO work and ISPS Code requirements on port security.

[30]WCO, Framework of Standards, § 1.3.

[31]The European Community's Customs Security Program, whose main element is the "authorized economic operator" concept, simply underlines that legitimate concern about container security on the part of all major trading nations.

manufacture and initial loading and forwarding. Applying the broad definition of MDA that underpins the National Strategy for Maritime Security (NSMS),[32] the complete supply chain might well be viewed as properly within the MDA focus. Of course, such a comprehensive approach raises questions about the allocation of responsibilities among competing organizations or agencies entrusted with maritime security—in short, functional specialization. Given that the present study's focus is MSP involving the traditional international maritime community—that is, navies, the maritime law enforcement community, and the shipping and fishing industries—and consistent with the emerging division of labor among international organizations,[33] only those shoreside segments of supply chain security that are located within or tied to the port of departure itself[34] are discussed in this appendix.

In sum, two cargo and container security measures are directly and to a large degree associated with MDA, yet also best handled by agencies traditionally concerned with the security of vessels and ports:

- The routine provision of cargo/container information to the destination state before the cargo or container is loaded on the vessel in the foreign port and
- Outbound security inspections of cargo and containers at the request of the destination state.

In imposing a 24-hour-in-advance electronic notification rule in 2002, the United States triggered the emergence of similar international normative expectations. Thus Standard 6 of the WCO's Framework of Standards recommends that customs administrations "require advance electronic information on cargo and container shipments in time for adequate risk assessment to take place."[35] Following this lead, the European Community now mandates a prearrival summary

[32]See supra note 1.

[33]The issue has been a matter of concern for the IMO. At its 34th session in March 2007, IMO's Facilitation Committee (FAL) approved a draft Joint Maritime Safety Committee/FAL circular on securing and facilitating international trade, which notes that the WCO has primacy over supply chain security, with IMO's role being limited to those container security aspects related to ships and port facilities. Similarly, on the domestic front, CBP has taken the lead role in cargo security. Only when cargo is moved on the waterborne leg of the trade route does USCG have oversight of the cargo's carriage requirements and the care needed for that cargo while on the vessel and at the port facility.

[34]The security screening and certification of cargo inland, that is, during its manufacture, loading, and transport into port, as well as of the parties involved in this process directly or indirectly—brokers, manufacturers, warehouse operators and carriers—represents a security function on the landward side of shipping operations.

[35]WCO Framework of Standards.

declaration for containerized cargo "at least 24 hours before loading at the port of departure."[36]

Similarly, the CSI has paved the way for the adoption by the WCO of Standard 11 of the Framework of Standards, which recommends that a state's "customs administration should conduct outbound security inspection of high-risk containers and cargo at the reasonable request of the importing country." Although the standard is formulated as a mere recommendation, requested states might legally be required to respond pursuant to the terms of any applicable bilateral customs mutual assistance agreement.[37]

Considering the worldwide interdependence of maritime cargo operations, their potential vulnerability to acts of terrorism, and the likely worldwide repercussions of a major breach of container security, it would seem prudent for the DHS (that is, for the CBP and USCG) and the DoS to support efforts to make advance electronic cargo reporting a general international legal requirement. Specifically, the 24-hour rule ought to become the binding legal standard applicable globally to international movements of closed cargo units.

Similarly, accession by the authorities at the port of departure to a request by the destination state for an outbound security inspection can be made mandatory as a matter of general international law and subject to safeguards regarding the reasonableness of such a request and legitimate expectations of privacy.

In parallel with these efforts, it would seem practical for CBP and other relevant agencies to extend the existing CSI program beyond its present geographic scope to additional foreign ports of interest.

Port and Coastal States' Maritime Domain Awareness

"An effective understanding of anything associated with the maritime domain" that carries national security implications[38] suggests an adequate, accurate, and timely flow of information to the actor(s) concerned. Viewed from an international legal perspective, MDA could be improved by expanding the present rights of port and coastal states to information about vessel movements; by improving general vessel tracking capabilities; and by strengthening maritime information exchanges through the expansion and better integration of global databases on

[36]This requirement is effective from July 1, 2009, and applies to containerized cargo only. See Art. 184a, § 1(a), of Commission Regulation (EC) No. 1875/2006 of December 18, 2006, amending Regulation (EEC) No. 2454/93 laying down provisions for the implementation of Council Regulation (EEC) No. 2913/92 establishing the Community Customs Code.

[37]For example, as of May 2007, the United States had entered into customs mutual assistance agreements with 60 countries and Taiwan. These agreements follow the WCO's model bilateral assistance agreement. The (multilateral) International Convention on Mutual Administrative Assistance in Customs Matters, the Johannesburg Convention, which was adopted in June 2003 but is not yet in force, expressly encourages such binding arrangements. See Arts. 10 and 48(2) of the Convention.

[38]See supra note 1.

vessels and ports. Some of these improvements, however, might involve changes in international law, a few of which might be sensitive.

Port State's Rights to Information

Since, generally speaking, access to a state's port is a privilege rather than a right, the port state[39] is entitled to set conditions for the entry of an incoming vessel, including the provision of information. Leaving aside situations covered by special agreement[40]—bilateral, regional, or global—pursuant to which a foreign flag vessel is granted the right of entry, a port state's freedom to regulate access to its ports will in theory thus be limited only by considerations of reciprocity and the state's obligation to provide international notification of any such requirements.

Apart from the 24-hour container rule, commercial vessels are at present also subject to port state requirements regarding notification of their arrival. Thus the United States has adopted a 96-hour notice of arrival (NOA) requirement.[41] Among member states of the EC, a somewhat less stringent standard of notification in advance of arrival applies: Incoming vessels must report either (1) at least 24 hours prior to arrival or (2) upon leaving the previous port, if the voyage is less than 24 hours or if the port of call is not known or changes during the voyage, as soon as this information becomes available.[42] More stringent reporting obligations apply to vessels coming from ports outside the EC and carrying dangerous or polluting goods.[43] The information to be communicated to the port state is of

[39]In an international legal sense, "port state" connotes a state that may have international jurisdiction over a foreign flag vessel on account of the vessel's declared intention to (voluntarily) visit that state's port. Reference to "coastal state" denotes a state that may have jurisdiction over a vessel with no intention to put into a port of that state on account of the fact that it transits the territorial sea (out to a distance of 12 miles from shore) or the exclusive economic zone (out to 200 miles) of that state. The flag state, finally, is the vessel's national state—i.e., the state whose flag the ship is entitled to fly. As a general rule, the flag state has primary jurisdiction over its vessel; however, in certain situations its jurisdiction may be concurrent only with that of the port or coastal state.

[40]Indeed, states frequently enter into bilateral or multilateral agreements as a result of which they are legally bound to open their ports to vessels flying the flag of a treaty party.

[41]See 33 CFR §160.212.

[42]See Art. 4, para.1, of Directive 2002/59/EC of the European Parliament and the Council of June 27, 2002, establishing a Community vessel traffic monitoring and information system and repealing Council Directive 93/75/EEC, OJ L 208/10, August 5, 2002.

[43]See Art. 4, para. 2, and Art. 13.

a general nature;[44] it is not security-specific as such but obviously will have some security implications.[45]

On top of the general prearrival notification, port states today are specifically authorized to obtain notification of certain security information prior to the vessel's entry into port.[46] These states are entitled to information on the vessel's security status pursuant to SOLAS Chapter XI-2 and the ISPS Code, specifically on whether the vessel carries a valid International Security Certificate, as required; the security level at which the vessel operates; the security level at which the vessel operated in any previous port where the vessel has conducted a ship/port interface; and so on.[47] Importantly, the vessel must keep a record of this information for the last 10 calls at port facilities,[48] which would be accessible to port authorities before the vessel's arrival in port, allowing the authorities to acquire a better picture of the security risks, if any, associated with the vessel.

Coastal State's Rights to Information and General Vessel Tracking

General international law, both customary law and UNCLOS, severely circumscribe a coastal state's informational rights regarding a vessel that—without intending to call at a port of that state—simply passes through its territorial sea or transits its exclusive economic zone (EEZ).[49] Thus it is generally agreed that a foreign flag vessel entering the territorial sea to exercise its right of innocent passage through these waters is not required to report its arrival to the coastal state or to provide other information such as its cargo. This is true, though the subject of some controversy, even for ships carrying "nuclear or inherently dangerous or noxious substances," which must carry documents and observe special

[44]Pursuant to 33 CFR §160.206, a notice of arrival must include information regarding vessel, voyage, cargo, crew, and persons on board, as well as vessel safety. The European Community law also prescribes submission of information on the vessel, voyage, cargo, crew, and any other person on board. See Annex I to Directive 2002/59/EC.

[45]For example, U.S. regulations require all vessels on an international voyage to provide information on their last five ports of call. Information-submitting parties must also provide a vessel's estimated time of departure and the name of the vessel charterer, if applicable.

[46]For the corresponding domestic regulations in the United States and the European Community, see 33 CFR, Part 104, and Art. 6 of Regulation (EC) No. 725/2004 of the European Parliament and the Council of March 31, 2004, on enhancing ship and port facility security, OJ L 129/6, April 29, 2004, respectively.

[47]See SOLAS Chapter XI-2; Special measures to enhance maritime security, Regulation 9, para. 2.1.

[48]See SOLAS Chapter XI-2; Special measures to enhance maritime security, Regulation 9, para. 2.3.

[49]Of course, to the extent that foreign flag vessels intend to engage in activities in the territorial sea or in the EEZ over which the coastal state has regulatory jurisdiction under UNCLOS, the vessel may be required to obtain permission from UNCLOS and a fortiori to notify the state of its intentions, including its arrival in the maritime zone concerned.

precautions established for such ships by international agreement.[50] Similarly, vessels navigating in the EEZ have no informational obligations to the coastal state. This appears, for example, to be true also of a fishing vessel entering the EEZ for the purpose of transiting the waters, not fishing therein, despite the fact that the coastal state enjoys sovereign rights to its EEZ's natural resources and therefore might be deemed to have a legitimate interest in being notified of the arrival or presence of foreign fishing vessels.

On the other hand, coastal states may be entitled to information bearing in particular on navigational safety, search and rescue (SAR), marine pollution prevention and control, and thereby also, albeit indirectly, on maritime security. For example, upon entry into areas of the sea subject to a mandatory Ship Reporting System (SRS),[51] a vessel must report to the appropriate coastal authority all required information in accordance with the provisions of the system. In general, the information to be supplied will be limited to the ship's name, call sign, IMO identification number, and position. However, information on any operational defects of the ship and the nature of its cargo, if hazardous, might have to be communicated as well to the coastal state authorities.[52] By the same token, vessels entering an area of vessel traffic services (Vessel Tracking System [VTS])[53] within a coastal state's territorial sea will generally be required to report to the coastal state authorities, usually by radio, and may be tracked by the VTS control center. The use of VTS may be mandated only in sea areas within the territorial sea of a coastal state.[54]

Similarly, IMO-approved mandatory ship routing systems,[55] including traffic separation schemes, deepwater routes, areas to be avoided, and the like, entail restrictions on vessel navigation, anchoring, and so on. These systems aim at enhancing maritime traffic safety and protecting the marine environment. A ves-

---

[50]See Art. 23 of UNCLOS.

[51]See SOLAS, Chapter V, Regulation 11 (mandatory if and when approved as such by IMO's MSC).

[52]See Guidance Note on the Preparation of Proposals on Ships' Routing Systems and Ship Reporting Systems for Submission to the Sub-Committee on Safety of Navigation, IMO Doc. Maritime Safety Committee Circ.1060, January 6, 2003, Annex, p. 5, para. 6.2.2.

[53]The purpose of a VTS is to provide active monitoring and navigational advice for vessels in particularly confined and busy waterways. There are two main applications of VTS: (1) systems subject to surveillance that involve one or more land-based sensors (radar, AIS, and closed circuit television sites) and (2) systems that output their signals to a central location where operators monitor and manage vessel traffic movement. Systems not subject to surveillance involve one or more reporting points at which ships are required to report their identity, course, speed, and other data to the monitoring authority.

[54]See SOLAS, Chapter V, Regulation 12, para. 3.

[55]See SOLAS, Chapter V, Regulation 10. Ship routing measures become mandatory if and when approved as such by IMO's Maritime Safety Committee.

sel that enters any such area must comply with applicable routing measures[56] and thus may be subject to monitoring by the coastal state(s) concerned.[57]

Limited though a coastal state's legal authority might be in obtaining information on or monitoring foreign flag vessels in the waters off its coast, this lack of authority is being offset by legal developments that have created or will create new vessel tracking capabilities. The first of these is the emergence of a general legal requirement that vessels be equipped with an Automatic Identification System (AIS).[58] The system,[59] which transmits a vessel's identifying signal and other relevant information,[60] could obviously be a potent tool for improving coastal states' MDA. However, AIS suffers from several drawbacks, including the distance range or frequency range over which its transmission can be received, the need for coastal state infrastructure, and the potential security and safety risks associated with open broadcasting of vessel data.

Effective January 1, 2008, a new IMO regulation on LRIT will enter into force as part of a revised SOLAS Chapter V: Safety of Navigation. It requires vessels subject to the regulation[61] to be fitted with LRIT equipment to automatically transmit information that will allow LRIT, both for security and SAR purposes, without unduly impacting the security of the transmitting vessel itself. For this reason, and unlike the fairly comprehensive vessel data made available through AIS, LRIT will divulge only the ship's identity, location, and date and time of its position. Moreover, there will be no interface between AIS and LRIT. Whereas AIS information is broadcast, hence potentially available to anyone, LRIT information will be available only to the flag state, the port state (if the vessel plans to call at its port(s)), governments conducting SAR operations and enquiring about ships in the area, and coastal states. The latter will be entitled to tracking informa-

---

[56]The Convention on the International Regulations for Preventing Collisions at Sea (COLREG), Regulation 10, also regulates the navigation of ships in or near traffic separation schemes established pursuant to SOLAS V/10.

[57]Ibid., Regulation 10, paras. 6 and 7.

[58]The AIS requirement is one of the results of the 2002 SOLAS Conference amending SOLAS Chapter V, Regulation 19. It applies to all ships of 300 GT or more engaged in international voyages and to ships of 500 GT or more not on international voyages, as well as all passenger ships irrespective of size. Although ostensibly a safety-related standard—AIS is part of SOLAS Chapter V focusing on navigational safety rather than of Chapter XI-2 dealing with special maritime security measures—it clearly has major security implications.

[59]For a full discussion of AIS's technical specifications and capabilities, see Chapter 3.

[60]The required data inputs, specified in IMO guidelines for the installation of shipborne automatic identification systems, include, inter alia, the vessel's position, heading, rate of turn, and navigational status. Additionally, information to be entered at initial installation of an AIS includes the maritime mobile service identity (MMSI) number, an IMO vessel number, the ship's name, its dimensions, and its type.

[61]The LRIT regulation applies to ships on international voyages: passenger ships, including high-speed craft; cargo ships, including high-speed craft, of 300 GT or more; and mobile offshore drilling units. The requirement will be gradually phased in after December 31, 2008.

tion from ships within 1,000 miles of their coasts, irrespective of whether or not the vessel intends to call at a port in the state concerned.[62]

Finally, fishing vessels are increasingly required to carry transmitters—vessel monitoring systems (VMSs)—that automatically report via satellite their positions at predetermined intervals or when requested. Moreover, VMS can deliver in near real time supplementary data on the vessel's catch, fishing activities, and so on. This growing mandatory use of VMS is explained by two developments: (1) a great portion of the high seas is now subject to a Regional Fisheries Management Organization (RFMO) scheme that may require fishing vessels to carry VMS as a condition of entry into that fishing area[63] and (2) a growing trend in the United States and EC countries[64] to stipulate that fishing vessels flying their flag be equipped with VMS, irrespective of where they fish.[65] This latter development has encouraged the expectation that in the near future the use of VMS on fishing vessels might become mandatory worldwide.[66] Clearly, a requirement for VMS on fishing vessels worldwide would significantly enhance MDA.

Enhanced Maritime Information Exchanges

Considering the importance of oceans for humankind as a whole,[67] it is not surprising that nowadays data are routinely being collected on every aspect of the state of the oceans, in particular the impact of human activities on it, including the operations of the vessels. In this vein, states have established several central-

[62]IMO's Web site claims that "the SOLAS regulation on LRIT does not create or affirm any new rights of States over ships beyond those existing in international law, particularly, the United Nations Convention on the Law of the Sea (UNCLOS), nor does it alter or affect the rights, jurisdiction, duties and obligations of States in connection with UNCLOS." Nevertheless, a coastal state's access to information on vessels outside its traditional jurisdictional reach is unprecedented.

[63]For example, Art. 11 of the North-East Atlantic Fisheries Commission's (NEAFC's) Scheme of Control and Enforcement (2007) requires each NEAFC contracting party to ensure that fishing vessels flying its flag carry VMS in the RFMO area for purposes of tracking the vessel and its catch.

[64]According to Commission Regulation (EC) No. 2244/2003 of December 18, 2003, which sets forth detailed provisions regarding satellite-based VMSs, all EC fishing vessels subject to VMS must have a satellite-tracking device installed on board to ensure automatic transmission to the Fishing Monitoring Center of the flag member state, at all times, of data relating to the fishing vessel's identification, its most recent geographical position, the date and time of the said position, and, effective January 1, 2006, its speed and course.

[65]A list of VMS programs worldwide can be found at <http://www.fao.org/fishery/vms/3>. Accessed on April 17, 2008.

[66]Note that the Magnuson-Stevens Fishery Conservation and Management Reauthorization Act, 2006, authorizes the Secretary of Commerce to support coordinated international efforts to ensure that all large-scale fishing vessels on the high seas be equipped with monitoring systems by December 31, 2008.

[67]For example, the U.S. Commission on Ocean Policy, in *An Ocean Blueprint for the 21st Century: Final Report* 1-2 (2004), points to the ocean as a highway for transporting goods and people; as a source of food, energy, and, potentially, life-saving drugs; as a venue for recreation and tourism; as a regulator of global climate; and as a cultural asset and source of aesthetic pleasure.

ized international databases to track compliance by flag states, vessel operators and owners, and other relevant actors with applicable international regulatory standards related to the environment, fisheries protection, navigational safety, or maritime security. Most of these maritime information exchange systems, listed in Appendix F, are at least potentially useful for MDA. If these information exchanges were to cooperate by sharing information on vessels, cargoes, and operators of interest, their aggregate value to maritime security could be considerable.

Among security-specific databases, mention must be made of the Global Integrated Shipping Information System (GISIS), a Web-based data system that permits verification of compliance with the maritime security provisions of SOLAS Chapter XI-2 and the ISPS Code. However, the GISIS security-related information is limited and includes only data that SOLAS contracting states must provide pursuant to SOLAS Chapter XI-2/13.[68] On the other hand, detailed data on vessels are being collected by port state information exchanges, the premier example of which is the Equasis database. This database involves all states participating in the Paris and Tokyo Memorandum of Understanding on Port State Control, the European Maritime Safety Agency, the USCG, and other maritime organizations and entities. GISIS offers a compilation of information on merchant vessels over 100 GT, which is generated in the course of port state inspections typically checking on vessel compliance with international standards on marine environmental protection, maritime safety, seafarers' well-being, fisheries protection, and maritime security. There is additional input from various private data providers, including classification societies, the P&I Clubs, and Lloyd's of London. The EC maintains a similar system for exchanging information, *SafeSeaNet*. However, that database is primarily geared to tracking vessel operations in European waters from a safety and pollution prevention perspective.

Other maritime-security-relevant multilateral databases include the proposed Central Automated (Cargo/Customs) Information System, which would facilitate information exchanges between national customs authorities on container security. Additionally, there exist some regional arrangements, a prime example of which is the Cooperating Nations Information Exchange System (CNIES), described in greater detail in Chapter 3.

One of the major blemishes on the global maritime information picture is the fact that there is still no centralized, comprehensive, and reliable database on high-seas fishing vessels, let alone a database that covers all fishing vessels above a certain minimum size. FAO maintains a database, the High Seas Vessel Authorization Record. However, not only is input into that data system incomplete and

[68]This includes, inter alia, national contact details, approved port facility security plans, and any changes thereto.

sporadic because only a limited number of states participate, but it is exclusively flag-state-based and therefore suffers from a critical systemic flaw.[69]

Specific Observations on the Port and Coastal States' Informational Rights, Vessel Tracking, and Maritime Exchanges

Like port states, states in general make access to port contingent on a vessel's prior notification of its arrival. Notice-of-arrival (NOA) requirements tend to vary from country to country, although the EC has established a common 24-hour rule for the ports of its member states. The United States subjects incoming vessels to a more stringent 96-hour standard. A port state's ultimate sanction for noncompliance with its NOA requirement is denial of entry into port.

By virtue of general international law as well as maritime-security-specific special legislation—namely, SOLAS Chapter XI-2—port states are legally in a position to demand security-relevant information from any vessel in advance of its arrival in port. Once a vessel is in port, the port state may be able to secure additional security-relevant information:

- Since pursuant to SOLAS Chapter XI-1, on special measures to enhance maritime safety, any foreign flag vessel is subject to port state control on operational requirements, port authorities will have access to the vessel's continuous synopsis record (CSR). The CSR lists details of the history of the ship, such as its registration, ownership, charter status (if applicable), classification, and so on[70]—in short, information with significant security implications.
- Exceptionally, a port state might be able to also gain access to the voyage data recorder (VDR) that passenger ships and ships other than passenger ships of 3,000 GT or more installed on or after July 1, 2002. Although the primary purpose of a VDR is to assist in accident investigations, it could help in reconstructing a suspect vessel's operational status and movements prior to its arrival at the port.[71]

Under the general law of the sea, a coastal state's right to information on vessels that merely pass through or transit its offshore waters (the territorial sea and the EEZ) are limited. However, to the extent that any portion of its territorial sea is subject to a special IMO-approved mandatory regime for the purpose of

[69]The critical weakness is that a system such as HSVAR relies on "the authenticity of information provided by or through the flag State of the vessel concerned." (See High Seas Task Force, "How to Get Better Information about High Seas Fishing Vessels," HSTF/05, February 25, 2005, p. 2.)

[70]See SOLAS Chapter XI-1, Regulation 5, para. 3.

[71]In December 2004, IMO's Maritime Safety Committee adopted amendments to SOLAS Chapter V, Regulation 20, on a phased-in carriage requirement for a shipborne simplified voyage data recorder (S-VDR). The amendments entered into force on July 1, 2006. This S-VDR still requires secure and retrievable storage of information concerning the position, movement, physical status, and command and control of a vessel during the period leading up to and following an incident.

vessel traffic management or marine environmental protection, such as the VTS, the SRS, or a ship-routing system, the coastal state will be legally entitled to information on vessels traversing these waters, information that may also have value from a security perspective.

Additionally, a coastal state may have the right to obtain information from vessels physically present in the territorial sea, the contiguous zone,[72] or the EEZ in the course of enforcing its laws vis-à-vis a vessel suspected of an infraction. However, this right presupposes that, as a matter of international law as set out in various provisions of UNCLOS, the coastal state's law and regulations on customs, fiscal, sanitary, and immigration matters; fisheries; and marine environmental protection actually do apply to the maritime zone in which the alleged infraction occurred and are enforceable given the location of the vessel when challenged by the coastal state's law enforcement agency.[73] Although the information a vessel would have to provide in these circumstances[74] would be directly related to the suspected infraction (and there is, generally speaking, no coastal state jurisdiction regarding maritime security offenses[75]), it stands to reason that such information might also be useful from a security perspective.

Finally, coastal states' MDA is being given a potentially substantial boost by the improvement of global vessel tracking capabilities, based on current AIS carriage requirements and the soon-to-be-operational LRIT information system.

- One of the acknowledged shortcomings of both AIS and LRIT is that their threshold of application—vessels of 300 GT or above—is relatively high and tends to exclude many vessels of interest from a maritime security standpoint.
- The utility of AIS data and LRIT information could be maximized if they were collated and more widely distributed among maritime security decision makers. However, at present any proposal for integrating tracking data acquired

---

[72]States may claim a zone contiguous to the territorial sea, which usually extends to 12 miles from the shore, out to a maximum distance of 24 miles. In this zone the coastal state may exercise limited jurisdictional powers in relation to the infringement of customs, fiscal, immigration and sanitary laws, and regulations that apply to its territory and territorial sea.

[73]Without going into unnecessary details, UNCLOS regulates in complex fashion a coastal state's jurisdiction, an example of which is Art. 220, paras. 2-6, bearing on the enforcement by the coastal state of its laws for the prevention and control of marine pollution.

[74]Note, for example, UNCLOS, Art. 220, para. 3, which provides as follows: "Where there are clear grounds for believing that a vessel navigating in the exclusive economic zone or the territorial sea of a State has, in the exclusive economic zone, committed a violation of applicable international rules and standards for the prevention, reduction and control of pollution from vessels or laws and regulations of that State conforming and giving effect to such rules and standards, that State *may require the vessel to give information* regarding its identity and port of registry, its last and its next port of call and other relevant information required to establish whether a violation has occurred" (emphasis added).

[75]Except perhaps for certain violations of the coastal state's criminal law by a vessel in that state's territorial sea, as discussed below in the section "Vessel Boarding."

through AIS or LRIT runs afoul of concerns about access to data, privacy, and so on.[76]

Today there are a number of international, multilateral maritime information exchange systems (see Appendix D) of differing quality and relevance to maritime security, the most promising being the Equasis database. Some of these systems are up and running, and others are still being set up. All of them, whether specifically dedicated to maritime security or not, presently do collect or eventually will collect, collate, and store data of potentially significant value from a security perspective.

These information systems could provide an integrated data platform, which from the perspective of enhancing maritime security would be more valuable than the sum of its parts. The political challenge therefore will be to persuade states, the maritime community, and civil society that it might be possible to improve security through data sharing across these systems without sacrificing legitimate privacy interests or abandoning other safeguards against potential abuse.

It would seem prudent for the DHS (the USCG), the DoS, and other U.S. stakeholders to support efforts to explore, within the IMO, the possibility of lowering the present 300 GT threshold for vessels subject to the AIS and the LRIT requirements. Also, it would be prudent for them to support the adoption of a global international legal requirement for electronic monitoring systems, such as the VMS, for fishing vessels.

All U.S. stakeholders, including the DOC's National Oceanic and Atmospheric Administration (NOAA), would benefit from a reliable system for monitoring fishing vessels worldwide, preferably as a stand-alone arrangement; the expansion of existing bilateral and multilateral cooperation based at the RFMO; and the integration of these databases into a global information system for fishing vessels. To this end, Equasis might serve as a model for an independent, multi-sourced, cost-effective international information system.

The United States has an interest in the wider sharing of maritime-security-relevant information presently held by international maritime information exchanges. All U.S. stakeholders would benefit by supporting the following efforts:

- Expansion of the reach of the Equasis database so that port state data from other regional port state control regimes that do not at present participate in Equasis could be fed into the system;
- Harmonization of international reporting formats and procedures to per-

[76]Note in this context that at its 79th session in December 2004, IMO's Maritime Safety Committee warned "that the publication on the world-wide web or elsewhere of AIS data transmitted by ships could be detrimental to the safety and security of ships and port facilities and was undermining the efforts of the Organization and its Member States to enhance the safety of navigation and security in the international maritime transport sector."

mit the sharing of information on vessels, operators, and cargo among existing international (and national) maritime information exchanges; and

• Clarification and redress of legitimate concerns about data protection and privacy, concerns that invariably arise in the context of sharing information because some countries are more sensitive than others about these issues.

Vessel Boarding: Interdiction

The right to board foreign flag[77] merchant ships[78] is a critical component of any MDA-enhancing regime, because boarding directly serves the acquisition and verification of maritime-security-related information. From an international legal perspective, the boarding of foreign flag vessels that are in, bound for, or departing from a port or the internal waters of the boarding state is relatively unproblematic: A state's jurisdiction over foreign vessels in port or in its internal waters is, after all, "necessarily exclusive and absolute."[79] The situation is different, however, when the vessel is simply passing through the state's territorial sea, transiting that or another state's EEZ, or navigating on the high seas proper.[80] In these circumstances a state's right to board the foreign flag vessel will generally depend on the flag state's consent or on specific boarding authority derived from UNCLOS or customary international law.

If the authorities of one state would like to board the vessel of another state to respond to security-related concerns about that vessel, most flag states would likely accede to any reasonable boarding request. Permission to board could be granted ad hoc or may have been given in advance by way of special agreement between the flag state and the requesting state. Today, a large number of states have concluded bilateral agreements that facilitate, if not authorize in advance, the boarding of foreign flag vessels. The United States, in particular, has successfully established a network of cooperative arrangements, such as the maritime counterdrug agreements with Caribbean and Central and South American states[81] and ship boarding agreements to interdict weapons of mass destruction

---

[77]Of course under international law and in international waters, states have an indisputable right to exert jurisdiction over their nationals—that is, over vessels flying their flag.

[78]Warships and government vessels operated for noncommercial purposes enjoy immunity from other states' exercise of jurisdiction, including immunity from being boarded anywhere. In consequence, the following comments will address issues exclusively related to the boarding of foreign flag merchant or government vessels operated for commercial purposes.

[79]C.J. Marshall, in *The Schooner Exchange v. McFadden*, Supreme Court of the United States, 1812; 11 U.S. (7 Cranch) 116, 3 Led. 287.

[80]Thus Art. 86 of UNCLOS provides that its high seas provisions specifically apply to "all parts of the sea that are not included in the exclusive economic zone, in the territorial sea or the internal waters of a State, or in the archipelagic waters of an archipelagic State."

[81]The State Department's Bureau of International Narcotics and Law Enforcement Affairs lists 26 bilateral maritime counterdrug agreements between the U.S. and Caribbean and Central and South American nations. See <http://www.state.gov/p/inl/rls/nrcrpt/2007/vol1/html/80853.htm>. Accessed on April 17, 2008.

(WMD),[82] which give to varying degrees in advance (maybe 24 hours, maybe a month) permission for law enforcement agencies of the other cooperating party to board their flag vessels.

Such special treaty-based authorizations do not, of course, cover vessels of all nations of potential interest, nor do they always provide clear or unrestricted authority to the boarding state. Indeed, many flag states remain reluctant to enter into such arrangements or, if they do, will often make their consent to boarding subject to various conditions. As a result, boarding of foreign flag vessels suspected of posing a security risk may be legally difficult if not impossible, unless, of course, the risk is such that boarding and other action against the vessel would be justifiable under the doctrine of self-defense,[83] a situation not discussed further in this appendix. Instead, the focus here is on the limits of existing boarding authority under UNCLOS and customary international law, the absence thereof in other relevant maritime-security-related multilateral agreements, and the negative implications of this state of affairs for maximizing MDA.

Under general international law, a state's right to visit a foreign flag vessel without the consent of the flag state is generally a function of the location of the vessel or the maritime zone in which the vessel is being approached;[84] its status—stateless vessel, merchant vessel, or warship; and the specific activity the vessel is suspected of engaging in. Leaving aside a state's boarding authority derived from its status as a port state,[85] a coastal state has limited boarding authority over vessels passing through its territorial sea or archipelagic waters.[86] It may, of course, take action, including boarding foreign flag vessels, to prevent passage that is not innocent. Art. 19, para. 1, of UNCLOS defines passage as innocent as long as it is not prejudicial to the peace, good order, or security of the coastal state. However, it has been the long-standing position of the United

[82]Thus far the United States has concluded seven ship-boarding agreements modeled after its counterdrug agreements with various Caribbean and Latin American countries to "operationalize" PSI. The seven countries are Belize, Croatia, Cyprus, Liberia, Malta, the Marshall Islands, and Panama.

[83]Conservatively speaking, boarding would have to meet the classic test annunciated in the context of the *Caroline* incident—namely, that the necessity giving rise to the claim of self-defense is "instant, overwhelming, and leaving no choice of means, and no moment for deliberation." See Letter of Secretary of State Daniel Webster to Lord Ashburton, dated August 6, 1842; reproduced at the Avalon Project, Yale Law School. Available at <http://www.yale.edu/lawweb/avalon/diplomacy/britain/br-1842d.htm#ash1>. Accessed on April 17, 2008.

[84]The right of approach on the part of a warship (or other properly marked government vessel authorized to carry out law enforcement functions) implies the right to request information about a foreign vessel's identity, especially its nationality. This right does not per se imply also the right to visit or board (search, and so on) the foreign flag vessel. The right of approach can be exercised by a state's warship and analogous government vessels on the high seas (including, for these purposes, other states' EEZs) and the state's territorial sea and EEZ.

[85]This authority can be invoked not only in respect of a vessel in port or the internal waters, but also in respect of inbound or outbound vessels in the state's territorial sea. See UNCLOS, Art. 25, para. 2, and Art. 27, para. 2.

[86]Foreign flag vessels transiting archipelagic waters—except for specifically designated "archipelagic sea lanes" to which the transit passage regime applies—are subject to the same innocent passage regime that applies to the territorial sea.

States and other countries that "any determination of non-innocence of passage by a transiting ship must be made on the basis of the acts it commits while in the territorial sea,"[87] as specified in the all-inclusive list of Art. 19, para. 2. Thus a foreign flag vessel's mere carriage of controversial cargo (say, of components of WMD) will not, without the commission of an act, permit the coastal state to characterize the vessel's passage as noninnocent. Nor should the vessel's means of propulsion, flag, origin, destination, or purpose of voyage provide legitimate grounds for determining passage as noninnocent.[88] On the other hand, the coastal state is entitled to apply—indeed, may be required to enforce—its criminal laws regarding activities in its territorial sea. Thus in accordance with the requirements of United Nations Security Council Resolution 1540, states must adopt and enforce laws that prohibit any nonstate actor from manufacturing, acquiring, possessing, developing, transporting, transferring, or using WMD or that prohibit their means of delivery, and so on. To the extent that a coastal state's criminal law implementing Security Council Resolution 1540 covers the maritime transport of WMD and their components through its territorial sea,[89] a foreign flag carrier would be deemed to violate that state's laws, which might trigger the latter's enforcement jurisdiction, including the right to board.[90]

A coastal state's jurisdiction over foreign vessels further offshore—that is, within the contiguous zone (CZ) (up to 24 miles) or the EEZ (out to 200 miles)—is more attenuated still. The state's boarding authority is correspondingly limited, in the sense of being functionally restricted, to the enforcement of applicable laws bearing on certain coastal states' interests for the protection of which the respective maritime zones were established in the first place. Thus within the CZ a coastal state will be authorized to board a foreign flag vessel as part of its right to prevent violations of its customs, fiscal, immigration, and sanitary laws and regulations.[91] In the EEZ, the coastal state has the right to board foreign fishing vessels as one of a number of specifically authorized measures to ensure compliance with its natural resource management/fisheries protection laws.[92] Since the coastal state enjoys at least equivalent, if not stronger, rights regarding the management of natural resources or fisheries protection in its ter-

---

[87] See Message from the President of the United States Transmitting the United Nations Convention on the Law of the Sea, Senate Doc. 103-39, 103rd Congress, 2d Sess. (1994), p. 15.

[88] Ibid.

[89] For example, under U.S. law, trafficking in WMD is a criminal offense if committed in the United States or the special maritime and territorial jurisdiction of the United States (see 18 USC 39, § 831(c)). For purposes of federal criminal jurisdiction, the territorial sea of the United States (extending out to 12 miles offshore) is within the "special maritime and territorial jurisdiction of the U.S." within the meaning of title 18, USC.

[90] Thus Art. 27, para. 1(b), of UNCLOS specifically recognizes a coastal state's right to board a foreign flag vessel "in connection with a crime committed during its passage" provided "the crime is of a kind to disturb the peace of the country or the good order of the territorial sea. . . ."

[91] See UNCLOS, Art. 33. By the same token, the coastal state has the right to punish infringements of these laws and regulations as may have occurred in its territory or the territorial sea.

[92] UNCLOS, Art. 73, para. 1.

ritorial sea, fisheries-enforcement-related boarding authority exists a fortiori also in the territorial sea. UNCLOS bestows significantly more circumscribed boarding authority in relation to vessels navigating in the EEZ or territorial sea and suspected of having committed a violation—in the EEZ—of applicable pollution laws resulting in a substantial discharge.[93] Again, in relation to vessels closer to the coast—namely, navigating in the territorial sea and suspected of having violated therein applicable pollution laws—the coastal state has somewhat broader authority to board the vessel for the purpose of verifying the violation, securing evidence, and so on.[94]

Because ships on the high seas are subject to the exclusive jurisdiction of the flag state, they are not subject to boarding by any other state without the consent of the former, except in limited circumstances that have traditionally been recognized as entailing a right of visit or of boarding—that is, where there are reasonable grounds for suspecting the vessel to be engaged in piracy, the slave trade, or unauthorized radio broadcasting or to be a vessel without nationality.[95] The latter is directly related to the right of approach, which refers to a warship's general authority to request information from any foreign vessel anywhere at sea except in another state's territorial sea, for the purpose of verifying the vessel's identity (registration and nationality). If suspicions about the vessel's identity cannot be resolved by way of radio communications or, for example, the transmission by fax of relevant documentation or other such, the right of approach gives way to a right of visit, and the warship can proceed to a physical inspection on board.[96] If, at the end of this process, the vessel turns out to be stateless[97]or can be assimilated to being a stateless vessel,[98] it will be subject to the boarding state's laws and regulations as if it were a national vessel of that state.

Beyond these traditional legal bases, there exists no equivalent boarding authority under general international law [99] vis-à-vis vessels suspected of engag-

[93]See UNCLOS, Art. 220, paras. 5-6. Additional conditions related to, for example, the associated environmental threat and the evidentiary threshold must be met before boarding will be permitted.

[94]UNCLOS, Art. 220, para. 2.

[95]UNCLOS, Art. 110, para. 1.

[96]UNCLOS, Art. 110, para. 2.

[97]For example, the U.S. Maritime Drug Law Enforcement Act, 46a USC 1903, para. 2, defines a vessel without nationality as including (a) a vessel aboard which the master or person in charge makes a claim of registry, which claim is denied by the flag nation whose registry is claimed; (b) any vessel aboard which the master or person in charge fails, upon request of an officer of the United States empowered to enforce applicable provisions of U.S. law, to make a claim of nationality or registry for that vessel; and (c) a vessel aboard which the master or person in charge makes a claim of registry and the claimed nation of registry does not affirmatively and unequivocally assert that the vessel is of its nationality.

[98]Thus in accordance with UNCLOS, Art. 92, para. 2, "[a] ship which sails under the flags of two or more States, using them according to convenience, may not claim any of the nationalities in question with respect to any other State, and may be assimilated to a ship without nationality."

[99]As regards boarding on the high seas, the provisions of UNCLOS simply reflect customary international law.

ing in the trafficking of drugs,[100] people, or weapons, including WMD. Nor is there as yet similar authority to board a foreign flag vessel on the high seas suspected of causing pollution of the marine environment or violating applicable fisheries laws.[101] In short, general international law does not offer boarding authority specifically couched in maritime-security terms. It is true, of course, that existing, mostly functionally defined boarding rights carry weight from a security perspective. Information that might be lawfully gathered in the course of a vessel boarding related to a coastal state's enforcement of pollution laws and regulations might well turn out to be useful also from a maritime-security perspective. Still, when it comes to maximizing MDA and enabling action based thereon, the international legal basis must be adjudged to be deficient. It has remained so despite various international legislative efforts to change the rules of the game by facilitating boarding, if not outright nonconsensual boarding, of foreign flag vessels to redress recognizable security risks associated with the vessels concerned.

The most concerted effort in this respect was the review of the 1988 Suppression of Unlawful Acts (SUA) Convention and its Protocol, which resulted in the adoption, in 2005, of a new Protocol that fundamentally revises the original instruments. Although maritime interdiction of WMD and, associated therewith, the easing of traditional restrictions on the boarding, inspection, and further "processing" of vessels, their crew, and cargo were key objectives of this amendment process, in the end, the states negotiating these revisions could not move beyond flag state consent as the fundamental organizing principle for handling boarding issues under the 2005 Protocol. Thus Article 8*bis* emphatically underlines that the boarding of a ship navigating "seaward of any State's territorial sea" is impermissible "without the express authorization of the flag state," even if that vessel or a person onboard is reasonably suspected of having been, being, or about to be involved in an act of terrorism involving WMDs.[102] The same article sets out certain options for the flag state that might be viewed as mitigating somewhat the harshness of the article's rejection of nonconsensual boarding in such dire circumstances: Upon becoming a party to the 2005 Protocol, any state may declare that with respect to vessels flying its flag it accepts the principle of presumptive

---

[100]Thus Art. 17 of the 1988 Vienna Convention, supra note 14, which provides the basic multilateral legal framework for interdiction of drug trafficking at sea, simply reflects recognition of the traditional of the principle of flag state consent.

[101]Some fishing vessels may, of course, be subject to boarding on the high seas under applicable regional or subregional fisheries management regimes or arrangements. But this authority can be invoked only between participating states and their fishing vessels or, more specifically, is premised on the flag state either (1) being a party to the RFMO concerned or to the 1995 United Nations Agreement for the Implementation of the Provisions of the United Nations Convention on the Law of the Sea of 10 December 1982 relating to the Conservation and Management of Straddling Fish Stocks and Highly Migratory Fish Stocks Agreement ["FSA"] or (2) otherwise accepting the terms of the RFMO.

[102]See Article 8*bis*, para. 5.

authorization if its authorities do not respond in timely fashion to another party's request for boarding.[103] Alternatively, it may give notice in advance of its authorization to board and search its vessel to determine whether a WMD-related offense has been, is being, or is about to be committed.[104] Evidently, these optional declarations by states are a far cry from specific and mandatory language that, coupled with necessary safeguards against abuse, either eliminates altogether the need to obtain the flag state's consent or establishes a legal presumption that boarding, search, seizure, and so on are authorized. Similarly, negotiations on Security Council Resolution 1540 may initially have aimed at changing the international rules on vessel boarding. However, once again, the U.S.-led effort fell short of moving states toward acceptance of the principle of nonconsensual boarding in situations involving the maritime trafficking of WMD by nonstate actors.[105] Finally, the PSI—which, unlike the 2005 SUA revisions and Security Council Resolution 1540, is not per se an international legislative initiative, nor as such capable of effecting international legal change—expressly commits participating states to ensure that their interdiction efforts be "consistent with their obligations under international law and frameworks."[106]

The physical act of boarding is, of course, only a first step in the exercise by the boarding state of jurisdiction over the foreign flag vessel. The international lawfulness of boarding as such does not automatically also permit any conclusions about what additional steps the boarding state might be permitted to take in relation to the foreign vessel, its cargo, crew, or passengers. Rather, if there is boarding authority under general international law—say, on the grounds of reasonable suspicion that the vessel concerned is a pirate ship—the lawfulness of additional enforcement steps, such as search, seizure, arrest, and detention, will be a function of the specific reasons for which international law recognizes this exception to the flag state's exclusive jurisdiction: As a *hostis humani generis*, any pirate will be deemed fair game, hence legally subject to the full jurisdictional authority of the boarding state. Similarly, if boarding appears justified on account of reasonable suspicions about a vessel's identity, boarding will be strictly limited to permit a document check—i.e., to verify the vessel's true identity. By the same token, when a flag state consents to the boarding of its flag vessel by another state, whether by special agreement in advance or ad hoc, its permission may be limited to just that and not necessarily include also authorization to investigate or

[103]"Timely fashion" was elsewhere defined as 4 hours from the flag state's acknowledgment of the receipt of a request to confirm the vessel's nationality. See Art. 8*bis*, para. 5(d).

[104]Art. 8*bis*, para. 5(e).

[105]As Lars Olberg explains, one reason was "widespread concern about the resolution's origins in the U.S. desire to pull in support for the Proliferation Security Initiative (PSI) . . . [with] China, Russia, and many others making clear that this provision should not be understood as an authorization for interdictions not otherwise permitted by international law." See L. Olberg, 2006, "Implementing Resolution 1540: What the National Reports Indicate," *Disarmament Diplomacy,* 82, Spring, The Acronym Institute. Available at <http://www.acronym.org.uk/dd/dd82/82lo.htm>. Accessed on April 17, 2008.

[106]PSI, Principle 4.

to further "process" the vessel. Indeed, all multilateral and most bilateral boarding arrangements carefully spell out steps requiring matching flag state authorization, vessel boarding, search, and seizure.[107] Moreover, flag states may subject any or all of such segmented authorizations to conditions.

From the MSP perspective, facilitation of the boarding of foreign flag vessels in maritime areas beyond national jurisdiction or control, including establishment of a special security-related exception to nonconsensual boarding, might not be a panacea but would certainly be useful. Vessel boarding permits the acquisition and/or verification of relevant information and, equally important, represents a first step in the enforcement action continuum. An easing of existing international legal restrictions on foreign flag vessel boarding would serve as a deterrent.

Moreover, it is not clear to what extent a coastal state can exercise criminal jurisdiction over a foreign flag vessel during innocent passage simply on the grounds that the vessel carries WMD materials, components, and so on, and thereby violates the coastal state's laws. It is also not clear to what extent a state is entitled not only to stop and search such a vessel navigating in the contiguous zone but also to seize its WMD-related cargo, as the PSI Statement of Interdiction Principles[108] suggests.

Specific Observations

Thus far, the principle of exclusive flag state jurisdiction has survived all recent international legislative attempts to establish a separate exception for security-related boarding on the high seas. On the other hand, international treaty practice, in particular the bilateral practice of the United States, indicates an emerging trend toward facilitating boarding. There are several strands of this development. First, there is evidence of a progressive tightening of the deadline for flag states to respond to requests for boarding in international waters. Second, if by the end of this time period—usually 2 to 4 hours—the flag state cannot or will not respond to the request, its consent to boarding for purposes of document verification and/or search of the vessel will be presumed. Third, some treaties specifically require the flag state to "contract out" of the operating presumption of flag state consent. Finally, a number of agreements envisage the flag state's assignment to third states of all its rights under the agreement concerning suspect

[107] For example, Art. 17, para. 4, of the 1988 UN Convention Against Illicit Traffic in Narcotic Drugs and Psychotropic Substances differentiates carefully between flag state permission to stop and board the vessel, search the vessel, and—if evidence of involvement in illicit traffic is found—take appropriate action with respect to the vessel, persons, and cargo on board. Art. 9 of the Council of Europe Agreement on Illicit Traffic by Sea, Implementing Article 17 of the United Nations Convention Against Illicit Traffic in Narcotic Drugs and Psychotropic Substances, ETS No. 156, establishes various subcategories of permitted conduct, thereby further segmenting the boarding state's authority for "processing" the vessel, the cargo, and persons onboard.

[108] Principle 4(d) PSI Statement of Interdiction Principles. Signed on September 4, 2003, by Australia, France, Germany, Italy, Japan, The Netherlands, Poland, Portugal, Spain, the United Kingdom, and the United States.

vessels claiming the flag state's nationality.[109] By contrast, the 2005 revisions of the SUA Convention and Protocol retain flag state consent as the basic operating principle but permit states to declare in advance their consent to requests of boarding or their acceptance that in case of their failure to respond in timely fashion their consent may be presumed.[110] In the aggregate, then, the picture that emerges differs from agreement to agreement, with U.S. treaty practice itself showing considerable variation in the specific boarding rights the United States has obtained.

According to Art. 110 of UNCLOS, a vessel that fails to display a flag may be treated as a suspect vessel and boarded for purposes of verifiying documents bearing on its identity (nationality and registration). In practice, states in general may not be willing to authorize their warships or law enforcement vessels to board a foreign-flag vessel on the high seas (or, equivalently, exercising its freedom of navigation elsewhere) simply to verify the vessel's identity. This may be due to concern about interfering with legitimate maritime trade and commerce or over potential liability for any loss or damage resulting from the boarding. In consequence, many countries insist that boarding for documentation verification be preceded by an attempt to contact the (alleged) flag state to obtain its express consent. Some countries, however, do permit boarding without requiring an initial attempt to contact the purported flag state.[111] Functionally, today the electronic signals a vessel must broadcast—the AIS and, soon, the LRIT system—are equivalent to physically displaying the flag of the national state. Therefore, the absence of such identifying transmissions, like the absence of a properly displayed flag, could be deemed prima facie evidence of a suspect vessel and could bring the vessel concerned within the ambit of UNCLOS, Art. 110, para. 2, which expressly authorizes warships to proceed to verify the vessel's identify by boarding and checking its documentation.

There does not appear to be an easing of existing international legal restrictions on foreign flag vessel boarding.

Boarding of foreign flag vessels in areas subject to limited or partial coastal

---

[109]See, for example, Art. II of the 2004 Amendment to the Supplementary Arrangement Between the Government of the United States of America and the Government of the Republic of Panama to the Arrangement Between the Government of the United States of America and the Government of Panama for Support and Assistance from the United States Coast Guard for the National Maritime Service of the Ministry of Government and Justice.

[110]See Art. 8*bis,* paras. 5(d) and 5(e), of the 2005 Protocol to the SUA Convention.

[111]See, for example, Australia's Customs Act 1901, sections 184A(9) and 185A. Similarly, some bilateral agreements permit boarding, without initial attempts at contacting the flag state, for purposes of document verification aboard a vessel claiming the nationality of one of the agreement states but not displaying the national flag, not displaying any marks of its registration or nationality, and claiming to have no documentation on board the ship. See Art. 4, para. 4, of the 2005 Agreement Between the Government of the United States of America and the Government of Belize Concerning Cooperation to Suppress the Proliferation of Weapons of Mass Destruction, Their Delivery Systems, and Related Materials by Sea; see also identical provisions in other counterproliferation ship-boarding agreements between the United States on the one hand and Liberia and the Marshall Islands on the other.

state jurisdiction, such as the EEZ, the CZ, or the territorial sea, raises several conceptual issues as a result of the adoption of Security Council Resolution 1540, which could reshape the traditional understanding of the import of some key provisions of UNCLOS.

The Chief of Naval Operations needs to encourage the DoS, DOJ, and DHS, among others, to strengthen bilateral and multilateral efforts to facilitate the boarding of foreign-flag vessels in international waters by shortening the requested flag state's response time, establishing presumptive consent, and delegation of bilaterally granted boarding authority to third states, among others.

It would seem prudent to seek legal clarifications through consultations, in particular with the states that are parties to UNCLOS, as to whether:

- A foreign flag vessel that does not have an AIS or an LRIT system on board provides reasonable grounds for questioning the vessel's ostensible or claimed nationality and registration and would therefore be subject to boarding on the high seas for the purpose of checking the vessel's documents;
- A coastal state can exercise criminal jurisdiction on board a foreign flag vessel exercising innocent passage through the state's territorial sea[112] if it carries WMD-related materials, people, etc. whose maritime transportation is subject to criminal sanctions in accordance with the Security Council resolution 1540;
- A flag state whose vessel is reasonably suspected of engaging in or being part of a terrorist plot or otherwise being used or guided by terrorists must give its consent to boarding or, conversely, can be presumed to give its consent; or
- A vessel navigating in the contiguous zone and suspected of trafficking in, for example, WMD components could be considered as violating the coastal state's customs laws within the meaning of UNCLOS, Art. 33.

The Chief of Naval Operations needs to work with the Secretary of the Navy to ensure that the DOD and DHS reaffirm the rules of engagement along the lines of internationally established and commonly used law enforcement concepts regarding the right to board, investigate, seize, arrest, detain, and prosecute and to use reasonable force against resisting vessels or crews.

Finally, in the long term, it would benefit the United States (and, in turn, DoS and DHS) to seek international support for a new maritime-security-related exception to exclusive flag state jurisdiction over vessels on the high seas or exercising freedom of navigation elsewhere. Such a result is unlikely to be accomplished by amending the 2005 revisions of the Suppression of Unlawful Acts Convention and Protocol; at present the only conceivable, though still controversial, option might be to secure a binding Security Council resolution to this effect.

---

[112]In accordance with UNCLOS, Art. 27, para. 1(b), on the ground that the crime concerned "is of a kind to disturb the peace of the country or the good order of the territorial sea."

D

Specific Reference Information

INTERPOL

Interpol is the world's largest international police organization, with a membership of most of the world's countries. It was founded in 1923 to facilitate cross-border police cooperation and to support organizations (both governmental and nongovernmental) whose mission is to prevent or combat international crime. Its objective is to facilitate international police cooperation even where diplomatic relations do not exist between particular countries.

Interpol, operating through each country's national crime bureau (FBI for the United States), gives law enforcement entities around the world instant access to its databases. Each Interpol member can in turn offer access to its databases on a consultative basis to groups such as border patrols or customs authorities by expanding existing multilateral agreements.

Interpol ensures its continuing existence by developing services and training at all levels of technical sophistication for its membership, using an established global operation in over 200 sites around the world. Interpol's databases and services ensure that police worldwide have access to the information and consultative and field support services they need to prevent and investigate crime.

THE IMO AND ITS AUTOMATED IDENTIFICATION SYSTEM AND LONG-RANGE IDENTIFICATION AND TRACKING SYSTEM

The International Maritime Organization (IMO) is a permanent global organization founded in 1948 and having members in approximately 167 countries. Members include all of the major coastal states and ship-owning nations. It also has as members governments and NGOs that recognize and adhere to the agreements of the IMO. The Automated Identification System (AIS) and the Long-Range Identification and Tracking (LRIT) system are two examples of the multilateral agreements facilitated by the IMO. The AIS (line of sight, 25 nmi) and the LRIT (out to 1,000 nmi from shore) system provide a global capability for monitoring the identification, location, track, and contents of ships on the seas.

IMO has established global multilateral agreements that all parties comply with if they wish to use the AIS and LRIT ship tracking systems. Under the IMO charter, once a majority of the membership has agreed to the terms and conditions of an agreement, all IMO members must abide by it. In this way, the IMO organizing body relied on the desire of member nations to maintain their membership to persuade them to ratify the agreement. This condition of membership has streamlined the review process for a number of international agreements, thereby benefiting the organization as a whole rather than getting caught up in an endless list of the concerns of individual members.

Finally, in its international role, the IMO commits itself to offer consultative services on demand to member nations. These services include, but are not limited to, data collection, the development of data standards, limited data sharing, support services, and rules enforcement.

LLOYD'S OF LONDON

For over 300 years Lloyd's of London has been managing risk for its clients in a number of markets, including maritime matters. It is a commercial organization that continues to find innovative ways of recognizing, quantifying, and managing business and environmental risks. Information collection and selective sharing are at the core of its business.

It does this through various contracts, or bilateral agreements, with its customers that quantify the terms and conditions of the particular situation. Its organization consists of more than 225 syndicates and brokers working cooperatively and continuously with clients to assess their risks and place the appropriate information and management tools in the marketplace to counteract unexpected circumstances.

Lloyd's has contract partners in nearly every country in the world and in more than 85 percent of the companies on the Dow Jones and Fortune 500. It is regarded as a commercial entity that has ties and operations with both government and commercial interests.

The involvement of Lloyd's in the Malacca Strait incident of 2005 is an example of the critical role of the Lloyd's network and other commercial maritime players in cooperative solutions to issues of maritime safety and security. It was the raising of insurance rates for shipping in the region due to piracy and terrorist activities that brought the security issue to the boiling point and led to sharpened interest among shippers and insurers acting with and pressuring the coastal states to improve regional cooperation in dealing with the security of the vessels at sea. That incident also stimulated investment from countries such as the United States, Australia, and Japan that allowed the infrastructure to be developed.

USNS *COMFORT* DEPLOYMENT

The deployment of the USNS *Comfort* to the Caribbean during the summer and fall of 2007 is an excellent example of cooperation and good will leading to bilateral agreements between nations that could improve maritime security. The deployment of *Comfort* for 4 months to 13 countries throughout the Caribbean basin was intended to provide medical, dental, and engineering assistance to the local populations as well as to train and share information with the health ministries in those countries. It followed on the heels of a similar highly successful deployment of the USNS *Mercy* to the Indonesian archipelago in 2006 with the same basic missions. It falls within the initial cooperation and bilateral agreement quadrant of the model (see Figure 2.1 in Chapter 2 of this report) and has great potential for increasing cooperation and information sharing in the future.

Planning conferences were held for agencies in Washington, D.C., and SOUTHCOM and for the appropriate government ministries and their militaries in the host nations. Details were coordinated, dates and locations were established, and levels of support, training, and involvement were offered to those countries based on local requirements and desires. The USNS *Comfort* was able to deal with the basic medical and dental problems faced by the people of the region. The ship's crew consisted of Navy, USCG, Air Force, and Air Guard medical personnel as well as representatives of the U.S. Public Health Service and NGOs.

Such deployments can strengthen existing relationships with people and can serve as building blocks for better cooperation and information sharing with the respective governments, improving the maritime security situation in the region.

PROLIFERATION SECURITY INITIATIVE

The Proliferation Security Initiative (PSI), established in 2003 in response to the fear that WMD might be spreading through covert transport of materials and delivery systems, represents a mixed form of intensive multilateral information sharing. PSI has no formal organizational structure or legal basis in an international treaty or UN convention. It relies on bilateral and multilateral agreements among a group of 15 to 20 core supporters and some broad declarations of support informally (and mainly secretly) pledged by a larger group of states, involving, in all, nearly 80 states. The core supporters subscribe to a Statement of PSI Principles, which includes a commitment to improve constraints at borders, ports, in the air, on land, and at sea by exploiting national legislation and an implied willingness to take action such as interdiction, in port or on the sea. PSI exercises have been held with groups of countries throughout the maritime commons; regular meetings and gaming among the core countries represent ongoing efforts at further consultation and the convergence of legislation and action. There also have been a number of (largely unpublicized) actions under PSI.

PSI is reinforced by bilateral arrangements on specific issues. The United States, for example, has concluded agreements with seven flag states supporting the U.S. right, after a formal request, to board ships on the high seas. Together with the fleets of the core states, the fleets of these seven account for more than 70 percent of the world's total. PSI is now an agreement within (although not referenced by) two broad UN Security Council resolutions: 1540 (against nuclear terrorism and proliferation) and 1718 (action against nuclear developments in North Korea). PSI also will gain status as support grows for relevant amendments to the IMO Convention for the Suppression of Unlawful Acts Against the Safety of Maritime Navigation.

PSI's information sharing is specific and purpose driven, and it generally takes place in limited bilateral or multilateral channels suitable for intelligence data. Actions are often military and involve a range of government agencies, including the military. States can agree to take action or not, depending on their own national laws and interests. A number of key countries (India, China, Indonesia, and South Korea) remain outside PSI.

GULF OF GUINEA INITIATIVE

The Gulf of Guinea (GoG) Initiative currently falls within the bilateral cooperation quadrant of the model (see Figure 2.1); however, the increased U.S. presence is leading to more cooperation and information sharing among 13 regional states and their European and U.S. partners. The GoG Initiative grew out of increased concern on the part of Africa, but also of the United States, NATO, and the European Union, about the violence and political, economic, and military instability resulting from the area's emerging status as a major oil exporter. Based largely on an intersecting set of military-to-military relationships and the proactive programs of EUCOM, the initiative includes various strategies to improve information and situational awareness but is principally aimed at developing national military and nonmilitary capacities to ensure maritime safety and security.

This geographical area features nations with immature governments, large ungoverned spaces, tremendous natural resources, militant violence and piracy, and limited ability to deal with the issues of maritime security. The legal framework for maritime law enforcement is inadequate or nonexistent in many of the countries, and pervasive corruption and weak governance detract from efforts to build and sustain security.

The underlying assumptions are that the countries in the region will commit to reducing corruption and graft, embrace a regional approach to maritime security, work on a legal framework for maritime law enforcement, and commit to improved maritime security. For the U.S. side of the GoG partnership, the assumptions are that U.S. agencies and departments (DoS, DOJ, DOC, USAID, FBI, and CDC) will engage in support of U.S. policy for Africa and this initiative. In addition, it is assumed that NGOs and industry (especially the oil producers) will eventually participate and partner in this effort.

Conferences, coordination sessions, and high-level visits to the countries have been part of the effort to bring various parties together. A GoG workshop on maritime safety and security was held in Ghana (March 2006) to address threats and vulnerabilities in the maritime domain. Conferences later in 2006 were attended by the ministers of 11 GoG nations as well as representatives from 5 EU members, government and military representatives from the United States, and members of regional and international organizations. The 11 GoG nations committed themselves to improving maritime awareness and enhancing regional cooperation through national, subregional, and international legal and regulatory frameworks. An action plan was developed with objectives to be met in the near term (12 months), the medium term (2 to 3 years), and the long term (more than 3 years) for improving MDA, regulatory frameworks, regional cooperation, and public awareness. In addition a commitment was made to strengthen regional and political will by establishing strategies for maritime safety and security (e.g., for individual and collective state access to AIS data) as well as linked national-level commissions for coordinating activities at sea and in ports.

These initiatives were linked to the current EUCOM strategy for mitigating the conditions that foster extremism, increasing partnerships for regional stability, and creating an environment favorable to the expansion of free market economies. The emphasis is on capacity building, with provisions for training, exercises, and the provision of needed equipment, infrastructure, and software support. On-site activities in the region are being carried out by EUCOM's proactive Naval Component Commander (C6F/CNE) and supported by Navy and USCG elements and other coalition naval units.

The GoG Initiative is a reminder of the difficulty of achieving objectives over time and reinforces the point that DoS and interagencies should lead the effort through EUCOM in accordance with the Theater Engagement Plan for the near term, transitioning to the new AFRICOM when the time is right. The Navy or the USCG may be tasked to execute elements of the plan through the naval component commander. This model might well be followed in varying degrees with the other COCOMs based on their own theater engagement plans and bilateral and multinational relationships with coalition partners.

MALACCA STRAIT SECURITY INITIATIVE

A growing number of Southeast Asian nations have formed a multilateral organization to develop and maintain a "comprehensive real-time regional sea situation picture" focused mainly on the Strait of Malacca and the Strait of Singapore. Headquartered in Singapore, they have dealt with information gathering (and sharing) and the technologies required, the issues of interdiction, and the national laws needed to support the effort. Like JIATF-S, the Malacca Strait Security Initiative (MSSI) stands as a model for what can and should be done worldwide.

The MSSI started with coordinated patrols by Singapore and Malaysia in 1992. It expanded to Indonesia (July 2004) and then Thailand (September 2005). As of September 2006, it had been expanded to eight countries (Cambodia, Japan, Laos, Singapore, Thailand, Philippines, Myanmar, and South Korea).

Out of these efforts, a supporting information infrastructure has emerged. There is the Vessel Traffic Information System, which receives inputs from the closed circuit television surveillance system, AIS transponders, and the Singapore Port Traffic Management System. Since January 1, 2007, all licensed powered harbor and pleasure craft are required to have the Harbour Craft Transponder System (HARTS), which feeds into the Port Operations Control Center. The Regional Maritime Information Exchange (ReMIX) is targeted at the WPNS Ops community. Information is exchanged on sea robberies and piracy incidents, missing or hijacked ships, vessels in distress, and other maritime incidents. It is an Internet, Web-browser-based platform. Access is via user ID and password. Once logged in, navies are free to upload or download any information they need. Information is shared with the Combined Enterprise Regional Information Exchange System (CENTRIXS), which operates in nearly 30 other countries throughout the world, including in the United States.

TSUNAMI RELIEF

An example of a multilateral emergency response is the ad hoc organization that grew out of the end-of-2006 tsunami in the Indian Ocean. Because the Navy had assets in the general area it was able to move quickly into the affected area (including areas of Thailand and Indonesia). Initially, transportation into the area was by U.S. Navy helicopters, which delivered medical supplies, water, food, personnel, and so on. Quickly Australia, India, and a number of other countries responded and an ad hoc organization developed.

In the organization that developed, Australia apparently volunteered to act as traffic cop to ensure the most effective allocation of available assets. One of the most important side benefits of the tsunami relief effort for the United States has been the increasingly favorable way that the United States is viewed by the citizens of the countries that were helped. That success resulted in the recent deployment of the USNS *Comfort* to the area. Participants include five countries and nongovernmental organizations.

UNITAS

United States–South American Allied Exercise (UNITAS) is a navy-to-navy (military-to-military)-generated exercise and training maritime program. This program started over 45 years ago as a bilateral engagement around the Caribbean and Latin America. The planning and engagement was at the direction of the U.S. Navy. As time passed the Marines and the USCG became regular participants, making this a truly maritime-centered event. At first, engagement consisted of Navy vessels making port calls, conducting onshore and underway training, and progressing from country to participating country. In the early 1990s, the organization became more multilateral. It is no longer expected or required that the United States lead or even plan all of the events. For instance, the Colombian Navy could lead a multination maritime event.

In the late 1990s, SOUTHCOM moved from Panama to Miami, Florida. The move included the setting up of a naval component for SOUTHCOM, and more structured and dedicated UNITAS engagements ensued. Engagements are now conducted regionally and are not necessarily run by the U.S. Navy, although they are planned by the U.S. naval component of SOUTHCOM.

The long-term commitment of U.S. maritime forces to the program has led to increased international cooperation and understanding. The U.S. Navy recognized that some USCG and Marine Corps competencies are critical for an integrated engagement program. UNITAS operations are now bilaterally initiated but act multilaterally in specific geographic areas.

RIMPAC

Rim of the Pacific (RIMPAC) is a bilaterally initiated training and exercise program in the Pacific, planned and led by the Navy. Countries are invited to participate off Pearl Harbor in multicountry task forces and exercises. Participation varies but has included Japan, Korea, Taiwan, Australia, New Zealand, the Philippines, and others. It remains U.S.-led and planned but fosters close working relationships and mutual understanding.

COOPERATION AND AFLOAT READINESS AND TRAINING

Pacific bilateral exercises outside Hawaii are conducted with the maritime forces of Pacific countries. The Cooperation and Afloat Readiness and Training (CARAT) exercise series believes that it is making the "1,000-ship Navy" vision a reality.[1] CARAT is now in its 13th year and has had partners such as Brunei, Malaysia, the Philippines, Singapore, and Thailand. It is a military-to-military, navy-to-navy engagement process. The 2007 exercise began in the Philippines

[1]LT Ed Early, USN. 2007. "CNO's Vision of 1,000-ship Navy Tested by CARAT Exercise," *Navy Newsstand,* June 26.

and ended in Thailand. While cooperation between the U.S. and host-nation ship and aircraft crews is crucial for CARAT, the exercise also involves Marine Corps and USCG personnel. The exercises are free-form. If a country wants to focus on maritime security, then that is what is done.

Information sharing is a central premise of the 1,000-ship Navy, and in this exercise (CENTRIX) it is used extensively to communicate quickly and effectively at sea and ashore. Vietnam was an observer during two phases of the exercise and is expected to become a full member in the years ahead. The goal is to see CARAT become more multinational because transnational problems are multinational in nature.

CONTAINER SECURITY INITIATIVE

The Container Security Initiative (CSI) is a post-9/11, bilateral cooperative initiative that began with Canada, Singapore, and the Netherlands. The initiative stations U.S. Customs personnel overseas to help ensure that containers loaded in overseas ports and destined for the United States are not tampered with. Loading can mean two things: cargo loaded into a container while in a port or, more often, containers sent to a port for further transfer. U.S. Customs inspectors carry out, in cooperation with their host country counterparts, preloading inspections in the foreign port.

The desire is for worldwide participation, resulting in more secure cargo shipment, reduced losses in the port, and increased integrity of the global supply chain. To date, 49 agreements have been entered into, with more anticipated in the future. Although the agreements are bilateral, when 49 countries are signatories the effect is a multilateral system of cargo security.

JOINT INTERAGENCY TASK FORCE-SOUTH

JIATF-S is an interesting hybrid organization. From the U.S. perspective it is a military command and a joint interagency task force that reports to SOUTHCOM. It has both homeland security and homeland defense responsibilities. This DOD command is uniquely led by a USCG officer, underlining the law enforcement nature of the command. It is staffed by all of the U.S. military services and many U.S. federal law enforcement agencies. It is also supported by a number of U.S. intelligence agencies.

The mission is to counter illicit trafficking, to promote cooperation on security, and to coordinate country team and partner nation initiatives. Since 9/11, the scope of the mission has been expanded to include other security concerns. Drug traffickers are now categorized as narcoterrorists. JIATF-S is in reality a joint international interagency task force. Twelve countries have posted liaison officers to JIATF-S. Specific rules for information sharing protect sensitive and classified information.

Bilateral agreements, ratified by the nations involved, now exist between numerous countries. These bilateral agreements were negotiated by DoS. Negotiation begins with an eight-part model counterdrug bilateral agreement. Most agreements differ one from the other because of the different viewpoints of the countries involved. These differences, combined with the particular requirements for information and intelligence sharing, create a complicated operational response structure that over time has become very successful. In the aggregate, JIATF-S operates multilaterally as it prosecutes counternarcotics cases and searches across the different national capacities for the right tools to deal with specific incidents.

The success of JIATF-S has increased steadily over time by strengthening intelligence and information gathering. At the same time the assets dedicated to the mission have steadily decreased. The single law-enforcement-centric mission made it easy for many nations to participate. The recent expansion of the mission scope of JIATF-S has raised concerns on the part of some partner nations, but to date has caused no adverse reaction. The significant trust that exists between partner nations did not occur overnight. Attention to national concerns and information sharing opened the doors to increased cooperation. Every contributor is a valued partner in the process, and information sharing is the goal.

E

Land Imaging Satellites

The five tables on pages 202-211 are reprinted as received, with permission from Noblis, Inc. ©2007, from "ASPRS Guide to Land Imaging Satellites," by W.E. Stoney of Mitretek Systems, updated for the NOAA Commercial Remote Sensing Symposium, "Key Trends and Challenges in the Global Marketplace," Washington, D.C., September 12-14, 2006. See <http://www.asprs.org/news/satellites/>. Accessed on October 25, 2007.

The two unnumbered graphs on pages 212 and 213 are reprinted as received, with permission from Noblis, Inc. ©2007, from "The Evolving World of Land Imaging Satellites: A GEOSS Opportunity," by W.E. Stoney of Mitretek Systems, presented to the GEOSS [Global Earth Observation System of Systems] Challenges and Opportunities Session at the IEEE IGARSS [International Geoscience and Remote Sensing Symposium], Barcelona, Spain, September 23, 2007.

CURRENT AND PLANNED, 36 M & BETTER, LAND IMAGING SATELLITES

SATELLITE	COUNTRY	LAUNCH	PAN RES. M	MS RES. M
Landsat 5	US	03/01/84		**30.0**
SPOT-2	France	01/22/90	**10.0**	20
ERS-2	ESA	04/21/95	**30.0**	
RadarSat 1	Canada	11/04/95	**8.5**	
IRS 1C	India	12/28/95	**6.0**	23
IRS 1D	India	09/29/97	**6.0**	23
SPOT-4	France	03/24/98	**10.0**	20
Landsat 7	US	04/15/99	**15.0**	30
IKONOS-2	*US*	09/24/99	**1.0**	4
TERRA (ASTER)	Japan/US	12/15/99		15, 30, 90
ARIRANG-1 (KOMPSAT-1)	Korea	12/20/99	**6.6**	
EO-1	US	11/21/00	**10.0**	30
EROS A1	*Israel*	12/05/00	**1.8**	
QuickBird-2	*US*	10/18/01	**0.6**	2.5
Proba	ESA	10/22/01	**8.0**	18, 36
ENVISAT	ESA	03/01/02	**30.0**	
SPOT-5	France	05/04/02	**2.5**	10
DMC AlSat-1 (SSTL)	Algeria	11/28/02		32
OrbView 3	*US*	06/26/03	**1.0**	4
DMC NigeriaSat-1 (SSTL)	Nigeria	09/27/03		32
DMC BilSat (SSTL)	Turkey	09/27/03	**12.0**	26
DMC UK (SSTL)	UK	09/27/03		32
IRS ResourceSat-1	India	10/17/03	**6.0**	6, 23, 56
CBERS-2	China/Brazil	10/21/03	**20.0**	20
FormoSat (RocSat2)	Taiwan	04/20/04	**2.0**	8
ThaiPhat (SSTL)	Thailand	12/01/04		36
IRS Cartosat 1	India	05/04/05	**2.5**	
MONITOR-E -1	Russia	08/26/05	8.0	20
Beijing-1 (SSTL)	China	10/27/05	**4.0**	32
TopSat (SSTL)	UK	10/27/05	**2.5**	5
ALOS	Japan	01/24/06	**2.5**	10
ALOS	Japan	01/24/06	**10.0**	
EROS B1	*Israel*	04/25/06	**0.7**	
Resurs DK-1 (01-N5)	Russia	06/15/06	**1.0**	3
Arirang-2 (KOMPSAT-2)	Korea	07/28/06	**1.0**	4

TerraSAR X	Germany	10/31/06	**1.0**	
RazakSat*	Malyasia	11/01/06	**2.5**	5
VinSat-1 (SSTL)	Vietnam	11/01/06	**4.0**	32
Sumbandilasat	South Africa	12/12/06		6.5
RadarSat 2	Canada	12/15/06	**3.0**	
RISAT	India	01/30/07	**3.0**	
IRS Cartosat 2	India	03/15/07	**1.0**	
GeoEye-1	*US*	03/16/07	0.41	1.64
RapidEye-A	*Germany*	06/01/07		6.5
RapidEye-B	*Germany*	06/01/07		6.5
RapidEye-C	*Germany*	06/01/07		6.5
RapidEye-D	*Germany*	06/01/07		6.5
RapidEye-E	*Germany*	06/01/07		6.5
CBERS-2B	China/Brazil	06/15/07	**20.0**	20
THOES	Thailand	06/30/07	**2.0**	15
HJ-1-A	China	07/01/07		30, 100H
HJ-1-B	China	07/01/07		30,150,300
WorldView -1	*US*	07/01/07	**0.5**	
COSMO-Skymed-1	Italy	11/12/07	**1.0**	
HJ-1-C	China	03/01/08		5, 20
EROS C	*Israel*	03/21/08	**0.7**	2.5
X-Sat	Singapore	04/16/08		10
CBERS-3	China/Brazil	05/01/08	**5.0**	20
COSMO-Skymed-2	Italy	05/01/08	**1.0**	
WorldView -2	*US*	07/01/08	**0.5**	1.8
Venus	Israel/France	08/01/08		5.3
TerraSAR L	Germany	08/15/08	**1.0**	
COSMO-Skymed-3	Italy	11/01/08	**1.0**	
Alsat-2A	Algeria	12/01/08	**2.5**	10
IRS ResourceSat-2	India	12/15/08	**6.0**	6, 23, 56
Pleiades-1	France	03/01/09	**0.7**	2.8
COSMO-Skymed-4	Italy	05/01/09	**1.0**	
TanDem-X	Germany	06/30/09	1.0	
Alsat-2B	Algeria	12/01/09	**2.5**	10
CBERS-4	China/Brazil	07/01/10	**5.0**	20
Spain Sat	Spain	07/01/10	**2.5**	
Pleiades-2	France	09/01/10	**0.7**	2.8
LDCM	US	07/01/11	**10.0**	30

Radar *Commercial* * Near Equatorial Orbit Revised 9/7/06

LAND IMAGING SATELLITES BY COUNTRY

SATELLITE	COUNTRY	LAUNCH	PAN RES. M	MS RES. M
DMC AlSat-1 (SSTL)	Algeria	11/28/02		32
Alsat-2A	Algeria	12/01/08	**2.5**	10
Alsat-2B	Algeria	12/01/09	**2.5**	10
RadarSat 1	Canada	11/04/95	**8.5**	
RadarSat 2	Canada	12/15/06	**3.0**	
Beijing-1 (SSTL)	China	10/27/05	**4.0**	32
HJ-1-A	China	07/01/07		30, 100H
HJ-1-B	China	07/01/07		30,150,300
HJ-1-C	China	03/01/08		5, 20
CBERS-2	China/Brazil	10/21/03	**20.0**	20
CBERS-2B	China/Brazil	06/15/07	**20.0**	20
CBERS-3	China/Brazil	05/01/08	**5.0**	20
CBERS-4	China/Brazil	07/01/10	**5.0**	20
ERS-2	ESA	04/21/95	**30.0**	
Proba	ESA	10/22/01	**8.0**	18, 36
ENVISAT	ESA	03/01/02	**30.0**	
SPOT-2	France	01/22/90	**10.0**	20
SPOT-4	France	03/24/98	**10.0**	20
SPOT-5	France	05/04/02	**2.5**	10
Pleiades-1	France	03/01/09	**0.7**	2.8
Pleiades-2	France	09/01/10	**0.7**	2.8
TerraSAR X	Germany	10/31/06	**1.0**	
RapidEye-A	*Germany*	06/01/07		6.5
RapidEye-B	*Germany*	06/01/07		6.5
RapidEye-C	*Germany*	06/01/07		6.5
RapidEye-D	*Germany*	06/01/07		6.5
RapidEye-E	*Germany*	06/01/07		6.5
TerraSAR L	Germany	08/15/08	**1.0**	
TanDem-X	Germany	06/30/09	1.0	
IRS 1C	India	12/28/95	**6.0**	23
IRS 1D	India	09/29/97	**6.0**	23
IRS ResourceSat-1	India	10/17/03	**6.0**	6, 23, 56
IRS Cartosat 1	India	05/04/05	**2.5**	
RISAT	India	01/30/07	**3.0**	
IRS Cartosat 2	India	03/15/07	**1.0**	
IRS ResourceSat-2	India	12/15/08	**6.0**	6, 23, 56

EROS A1	*Israel*	12/05/00	**1.8**	
EROS B1	*Israel*	04/25/06	**0.7**	
EROS C	*Israel*	03/21/08	**0.7**	2.5
Venus	Israel/France	08/01/08		5.3
COSMO-Skymed-1	Italy	11/12/07	**1.0**	
COSMO-Skymed-2	Italy	05/01/08	**1.0**	
COSMO-Skymed-3	Italy	11/01/08	**1.0**	
COSMO-Skymed-4	Italy	05/01/09	**1.0**	
ALOS	Japan	01/24/06	**2.5**	10
ALOS	Japan	01/24/06	**10.0**	
TERRA (ASTER)	Japan/US	12/15/99		15, 30, 90
ARIRANG-1 (KOMPSAT-1)	Korea	12/20/99	**6.6**	
Arirang-2 (KOMPSAT-2)	Korea	07/28/06	**1.0**	4
RazakSat*	Malyasia	11/01/06	**2.5**	5
DMC NigeriaSat-1 (SSTL)	Nigeria	09/27/03		32
MONITOR-E -1	Russia	08/26/05	8.0	20
Resurs DK-1 (01-N5)	Russia	06/15/06	**1.0**	3
X-Sat	Singapore	04/16/08		10
Sumbandilasat	South Africa	12/12/06		6.5
Spain Sat	Spain	07/01/10	**2.5**	
FormoSat (RocSat2)	Taiwan	04/20/04	**2.0**	8
ThaiPhat (SSTL)	Thailand	12/01/04		36
THOES	Thailand	06/30/07	**2.0**	15
DMC BilSat (SSTL)	Turkey	09/27/03	**12.0**	26
DMC UK (SSTL)	UK	09/27/03		32
TopSat (SSTL)	UK	10/27/05	**2.5**	5
Landsat 5	US	03/01/84		**30.0**
Landsat 7	US	04/15/99	**15.0**	30
IKONOS-2	*US*	09/24/99	**1.0**	4
EO-1	US	11/21/00	**10.0**	30
QuickBird-2	*US*	10/18/01	**0.6**	2.5
OrbView 3	*US*	06/26/03	**1.0**	4
GeoEye-1	*US*	03/16/07	0.41	1.64
WorldView -1	*US*	07/01/07	**0.5**	
WorldView -2	*US*	07/01/08	**0.5**	1.8
LDCM	US	07/01/11	**10.0**	30
VinSat-1 (SSTL)	Vietnam	11/01/06	**4.0**	32
Radar	*Commercial*	* Near Equatorial Orbit		Revised 9/7/06

OPTICAL LAND IMAGING SATELLITES BY BEST RESOLUTION

SATELLITE	COUNTRY	LAUNCH	PAN RES. M	MS RES. M	SWATH KM
	VERY HIGH RESOLUTION	**(.41 TO 1 METERS)**			
GeoEye-1	*US*	03/16/07	0.41	1.64	15
WorldView -1	*US*	07/01/07	**0.5**		16
WorldView -2	*US*	07/01/08	**0.5**	1.8	16
QuickBird-2	*US*	10/18/01	**0.6**	2.5	16
EROS B1	*Israel*	04/25/06	**0.7**		7
EROS C	*Israel*	03/21/08	**0.7**	2.5	16
Pleiades-1	France	07/01/08	**0.7**	2.8	20
Pleiades-2	France	07/01/09	**0.7**	2.8	20
IKONOS-2	*US*	09/24/99	**1.0**	4	11
OrbView 3	*US*	06/26/03	**1.0**	4	8
Resurs DK-1 (01-N5)	Russia	06/15/06	**1.0**	3	28
KOMPSAT-2	Korea	07/28/06	**1.0**	4	15
IRS Cartosat 2	India	08/15/06	**1.0**		10
	HIGH RESOLUTION	**(1.8 TO 2.5 METERS)**			
EROS A1	*Israel*	12/05/00	**1.8**		14
FormoSat (RocSat2)	Taiwan	04/20/04	**2.0**	8	24
THOES	Thailand	06/30/07	**2.0**	15	22, 90
SPOT-5	France	05/04/02	**2.5**	10	120
IRS Cartosat 1	India	05/04/05	**2.5**		30
TopSat (SSTL)	UK	10/27/05	**2.5**	5	10, 15
ALOS	Japan	01/24/06	**2.5**	10	35, 70
RazakSat*	Malyasia	11/01/06	**2.5**	5	?
Spain Sat	Spain	07/01/10	**2.5**		?
	HIGH MEDIUM RESOLUTION	**(4 TO 8 METERS)**			
Beijing-1 (SSTL)	China	10/27/05	**4.0**	32	600
VinSat-1 (SSTL)	Vietnam	11/01/06	**4.0**	32	600
MTI	US	03/12/00		5, 20	12

CBERS-3	China/Brazil	05/01/08	**5.0**	20	60, 120
CBERS-4	China/Brazil	07/01/10	**5.0**	20	60, 120
IRS 1C	India	12/28/95	**6.0**	23	70, 142
IRS 1D	India	09/29/97	**6.0**	23	70, 142
IRS ResourceSat-1	India	10/17/03	**6.0**	6, 23, 56	24, 140,740
IRS ResourceSat-2	India	12/15/06	**6.0**	6, 23, 56	24, 140, 740
RapidEye-A,B,C,D,E	*Germany*	06/01/07		6.5	78
KOMPSAT-1	Korea	12/20/99	**6.6**		17
R26m	South Africa	09/01/06		7.5	?
Proba	ESA	10/22/01	**8.0**	18, 36	14
MONITOR-E -1	Russia	08/26/05	8.0	20	94, 160
MEDIUM RESOLUTION (10 TO 20 METERS)					
SPOT-2	France	01/22/90	**10.0**	20	120
SPOT-4	France	03/24/98	**10.0**	20	120
EO-1	US	11/21/00	**10.0**	30	37
X-Sat	Singapore	04/16/08		10	50
LDCM	US	07/01/11	**10.0**	30	177
DMC BilSat (SSTL)	Turkey	09/27/03	**12.0**	26	24, 52
Landsat 7	US	04/15/99	**15.0**	30	185
TERRA (ASTER)	Japan/US	12/15/99		15, 30, 90	60
CBERS-2	China/Brazil	10/21/03	**20.0**	20	113
CBERS-2B	China/Brazil	06/15/07	**20.0**	20	113
LOW MEDIUM RESOLUTION (30 TO 56 METERS)					
Landsat 5	US	03/01/84		**30**	185
DMC AlSat-1 (SSTL)	Algeria	11/28/02		32	600
ThaiPhat (SSTL)	Thailand	12/01/04		36	600
DMC NigeriaSat-1 (SSTL)	Nigeria	09/27/03		32	600
DMC UK (SSTL)	UK	09/27/03		32	600
IRS ResourceSat-1 AWIFS	India	10/17/03		56	740
IRS ResourceSat-2 AWIFS	India	12/15/06		56	740

Commercial   * Near Equatorial Orbit   Revised 7/6/06

OPTICAL LAND IMAGING SATELLITES BY MS RESOLUTION

SATELLITE	COUNTRY	LAUNCH	PAN RES. M	MS RES. M	SWATH KM
HIGH RESOLUTION (1.64 TO 3 METERS)					
GeoEye-1	*US*	03/16/07	0.41	1.64	15
WorldView -2	*US*	07/01/08	**0.5**	1.8	16
QuickBird-2	*US*	10/18/01	**0.6**	2.5	16
EROS C	*Israel*	03/21/08	**0.7**	2.5	16
Pleiades-1	France	07/01/08	**0.7**	2.8	20
Pleiades-2	France	07/01/09	**0.7**	2.8	20
Resurs DK-1 (01-N5)	Russia	06/15/06	**1.0**	3	28
HIGH MEDIUM RESOLUTION (4 TO 8 METERS)					
IKONOS-2	*US*	09/24/99	**1.0**	4	11
OrbView 3	*US*	06/26/03	**1.0**	4	8
KOMPSAT-2	Korea	07/28/06	**1.0**	4	15
TopSat (SSTL)	UK	10/27/05	**2.5**	5	10, 15
RazakSat*	Malyasia	11/01/06	**2.5**	5	?
MTI VNIR	US	03/12/00		5	12
IRS ResourceSat-1 LISS-IV	India	10/17/03	**6.0**	6	24
IRS ResourceSat-2 LISS-IV	India	12/15/06	**6.0**	6	24
RapidEye-A,B,C,D,E	*Germany*	06/01/07		6.5	78
R26m	South Africa	09/01/06		7.5	?
FormoSat (RocSat2)	Taiwan	04/20/04	**2.0**	8	24
MEDIUM RESOLUTION (10 TO 26 METERS)					
SPOT-5	France	05/04/02	**2.5**	10	120
ALOS	Japan	01/24/06	**2.5**	10	35, 70
X-Sat	Singapore	04/16/08		10	50
THOES	Thailand	06/30/07	**2.0**	15	22, 90
TERRA (ASTER VNIR)	Japan/US	12/15/99		15	60

Proba	ESA	10/22/01	**8.0**	18	14
CBERS-3	China/Brazil	05/01/08	**5.0**	20	60, 120
CBERS-4	China/Brazil	07/01/10	**5.0**	20	60, 120
MONITOR-E -1	Russia	08/26/05	8.0	20	94, 160
SPOT-2	France	01/22/90	**10.0**	20	120
SPOT-4	France	03/24/98	**10.0**	20	120
CBERS-2	China/Brazil	10/21/03	**20.0**	20	113
CBERS-2B	China/Brazil	06/15/07	**20.0**	20	113
MTI VNIR TO TIR	US	03/12/00		20	12
IRS 1C	India	12/28/95	**6.0**	23	70, 142
IRS 1D	India	09/29/97	**6.0**	23	70, 142
IRS ResourceSat-1 LISS-III+	India	10/17/03		23	140
IRS ResourceSat-2 LISS-III+	India	12/15/06		23	140
DMC BilSat (SSTL)	Turkey	09/27/03	**12.0**	26	24, 52
	LOW MEDIUM RESOLUTION		**(30 T0 56 METERS)**		
EO-1	US	11/21/00	**10.0**	30	37
LDCM	US	07/01/11	**10.0**	30	177
Landsat 7	US	04/15/99	**15.0**	30	185
Landsat 5	US	03/01/84		**30**	185
TERRA (ASTER SWIR)	Japan/US	12/15/99		30	60
Beijing-1 (SSTL)	China	10/27/05	**4.0**	32	600
VinSat-1 (SSTL)	Vietnam	11/01/06	**4.0**	32	600
DMC AlSat-1 (SSTL)	Algeria	11/28/02		32	600
DMC NigeriaSat-1 (SSTL)	Nigeria	09/27/03		32	600
DMC UK (SSTL)	UK	09/27/03		32	600
ThaiPhat (SSTL)	Thailand	12/01/04		36	600
IRS ResourceSat-1 AWIFS	India	10/17/03		56	740
IRS ResourceSat-2 AWIFS	India	12/15/06		56	740
Radar	*Commercial*	10	* Near Equatorial Orbit		Revised 7/6/06

OPTICAL LAND IMAGING SATELLITES BY SWATH
LIMITED ANNUAL TOTAL GLOBAL COVERAGE CAPABILITY

SATELLITE	COUNTRY	LAUNCH	PAN RES. M	MS RES. M	SWATH KM
EROS B1	*Israel*	04/25/06	**0.7**		7
OrbView 3	*US*	06/26/03	**1.0**	4	8
IRS Cartosat 2	India	08/15/06	1.0		10
TopSat (SSTL) PAN	UK	10/27/05	**2.5**		10
IKONOS-2	*US*	09/24/99	**1.0**	4	11
MTI	US	03/12/00		5, 20	12
EROS A1	*Israel*	12/05/00	**1.8**		14
Proba	ESA	10/22/01	**8.0**	18, 36	14
KOMPSAT-2	Korea	07/28/06	1.0	4	15
TopSat (SSTL) MS	UK	10/27/05		5	15
GeoEye-1	*US*	03/16/07	0.41	1.64	15
WorldView -1	*US*	07/01/07	**0.5**		16
WorldView -2	*US*	07/01/08	**0.5**	1.8	16
QuickBird-2	*US*	10/18/01	**0.6**	2.5	16
EROS C	*Israel*	03/21/08	**0.7**	2.5	16
KOMPSAT-1	Korea	12/20/99	**6.6**		17
Pleiades-1	France	07/01/08	**0.7**	2.8	20
Pleiades-2	France	07/01/09	**0.7**	2.8	20
THOES PAN	Thailand	06/30/07	**2.0**		22
IRS ResourceSat-1 LISS-IV	India	10/17/03		**6.0**	24
IRS ResourceSat-2 LISS-IV	India	12/15/08		**6.0**	24
FormoSat (RocSat2)	Taiwan	04/20/04	**2.0**	8	24
DMC BilSat (SSTL) PAN	Turkey	09/27/03	**12.0**		24
Resurs DK-1 (01-N5)	Russia	06/15/06	**1.0**	3	28
IRS Cartosat 1	India	05/04/05	**2.5**		30
ALOS PAN	Japan	01/24/06	**2.5**		35
EO-1	US	11/21/00	**10.0**	30	37
X-Sat	Singapore	04/16/08		10	50
DMC BilSat (SSTL) MS	Turkey	09/27/03		26	52
CBERS-3 PAN	China/Brazil	05/01/08	**5.0**		60
CBERS-4 PAN	China/Brazil	07/01/10	**5.0**		60
TERRA (ASTER)	Japan/US	12/15/99		15, 30, 90	60

IRS ResourceSat-1 LISS-IV	India	10/17/03	**6.0**		70
IRS ResourceSat-2 LISS-IV	India	12/15/08	6.0		70
ALOS MS	Japan	01/24/06		10	70
IRS 1C PAN	India	12/28/95	**6.0**		70
IRS 1D PAN	India	09/29/97	**6.0**		70
THOES MS	Thailand	06/30/07		15	90
MONITOR-E -1 PAN	Russia	08/26/05	8.0		94

POTENTIAL FOR MULTIPLE MS CLOUD-FREE GLOBAL DATA SETS PER YEAR

CBERS-2	China/Brazil	10/21/03	**20.0**	20	113
CBERS-2B	China/Brazil	06/15/07	20.0	20	113
SPOT-5	France	05/04/02	**2.5**	10	120
CBERS-3 MS	China/Brazil	05/01/08		20	120
CBERS-4 MS	China/Brazil	07/01/10		20	120
SPOT-2	France	01/22/90	**10.0**	20	120
SPOT-4	France	03/24/98	**10.0**	20	120
IRS ResourceSat-1 LISS-III+	India	10/17/03		23	140
IRS ResourceSat-2 LISS-III+	India	12/15/08		23	140
IRS 1C MS	India	12/28/95		23	142
IRS 1D MS	India	09/29/97		23	142
MONITOR-E -1 MS	Russia	08/26/05		20	160
LDCM	US	07/01/11	10.0	30	177
Landsat 7	US	04/15/99	**15.0**	30	185
Landsat 5	US	03/01/84		**30**	185
RapidEye-A,B,C,D,E	*Germany*	06/01/07		6.5	390
Beijing-1 (SSTL)	China	10/27/05	**4.0**	32	600
VinSat-1 (SSTL)	Vietnam	11/01/06	4.0	32	600
DMC AlSat-1 (SSTL)	Algeria	11/28/02		32	600
ThaiPhat (SSTL)	Thailand	12/01/04		36	600
DMC NigeriaSat-1 (SSTL)	Nigeria	09/27/03		32	600
DMC UK (SSTL)	UK	09/27/03		32	600
IRS ResourceSat-1 AWIFS	India	10/17/03		56	740
IRS ResourceSat-2 AWIFS	India	12/15/08		56	740

Radar *Commercial* * Near Equatorial Orbit Revised 7/6/06

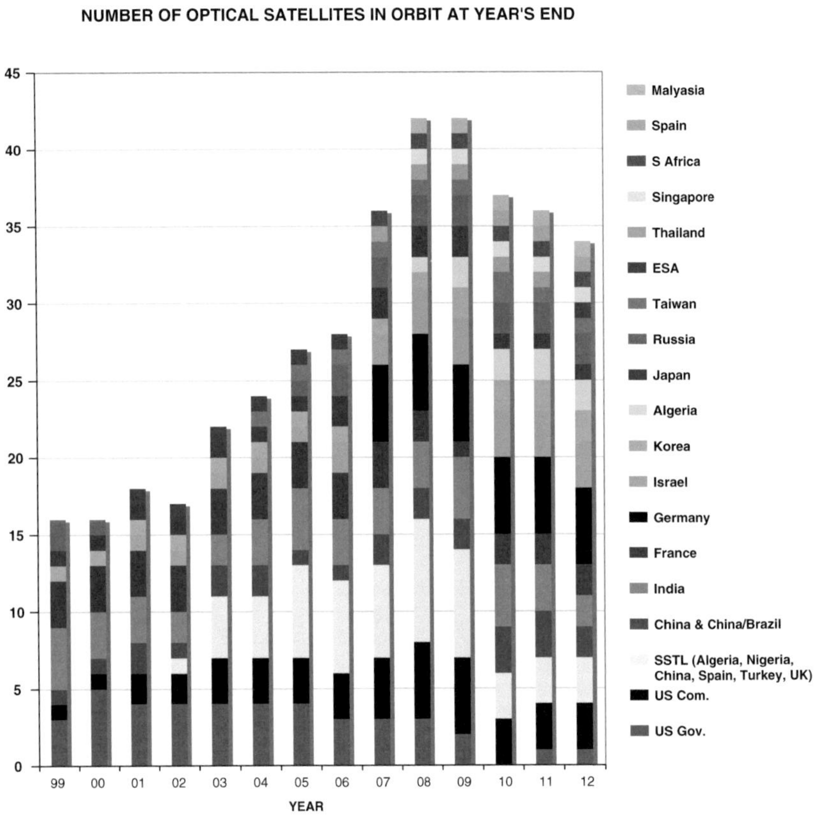

SOURCE: William E. Stoney, Mitretek Systems. 2007. "The Evolving World of Land Imaging Satellites: A GEOSS Opportunity," presented to the GEOSS [Global Earth Observation System of Systems] Challenges and Opportunities Session at the IEEE IGARSS [International Geoscience and Remote Sensing Symposium], Barcelona, Spain, September 23. Noblis, Inc. ©2007. Reprinted with permission.

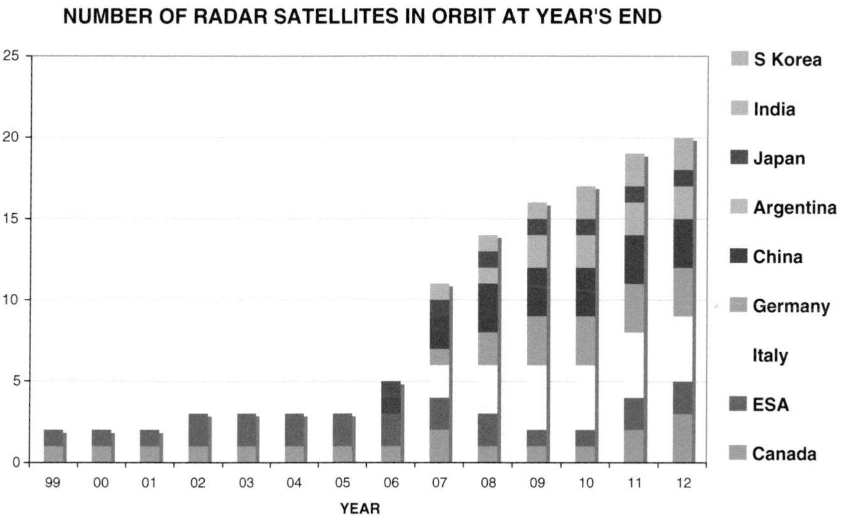

SOURCE: William E. Stoney, Mitretek Systems. 2007. "The Evolving World of Land Imaging Satellites: A GEOSS Opportunity," presented to the GEOSS [Global Earth Observation System of Systems] Challenges and Opportunities Session at the IEEE IGARSS [International Geoscience and Remote Sensing Symposium], Barcelona, Spain, September 23. Noblis, Inc. ©2007. Reprinted with permission.

F

International Databases as Potential Sources of Shared Information

GLOBAL INTEGRATED SHIPPING INFORMATION SYSTEM (GISIS)

- *Status:* Operational
- *Administration:* IMO, Web-based data system
- *Nature of information:* Maritime security-related. To permit verification of compliance with the maritime security provisions of SOLAS Chapter XI-2 and the ISPS Code
- *Scope of operation:* Covers all states parties to SOLAS
- *Legal basis:* Communication of information is legally required pursuant to SOLAS Regulation XI-2/13

PORT STATE INFORMATION EXCHANGES

Equasis

- *Status:* Operational
- *Administration:* Multilateral, public Web-based data system
- *Nature of information:* Primarily safety-related, port-state-relevant information provided by the following:

—States parties to the Paris and Tokyo Memorandum of Understanding on Port State Control and the USCG

—Classification societies

—P&I Clubs

—Lloyd's of London

—International Association of Independent Tanker Owners

—IMO

—European Maritime Safety Agency (EMSA)
—International Transport Forum
—Oil Companies International Marine Forum

- *Scope of operation:* Participation by the maritime administrations of Australia, France, Japan, Norway, Spain, the United Kingdom, and EMSA as full partners; IMO and USCG as observers
- *Legal basis:* (new) Memorandum of Understanding of 2007 (port state information itself is being gathered on the basis of various provisions of UNCLOS and other maritime conventions, such as SOLAS, ILO Convention No. 147, and the STCW Convention)

European Communities: SafeSeaNet

- *Status:* Operational
- *Administration:* Internet-based data exchange platform between the maritime administrations of EC member states
- *Nature of information:* Primarily safety- and pollution-prevention-related information on vessels in Europeans waters[1]
- *Scope of operation:* All EC states participate, as well as some non-EC states such as Norway
- *Legal basis:* EC Directive 2002/59/EC

INFORMATION ON FISHING VESSELS

High Seas Vessels Authorization Record (HSVAR)

- *Status:* Operational
- *Administration:* Food and Agriculture Organization of the United Nations
- *Nature of information:* Data on individual vessels authorized to fish on the high seas as means to counter illegal, unreported, and unregulated fishing
- *Scope of operation:* Global, applies to all fishing vessels on the high seas
- *Legal basis:* Article VI of the 1993 Agreement to Promote Compliance with International Conservation and Management Measures by Fishing Vessels on the High Seas (not yet in force)

[1]Covers ships of 300 GT and upward, unless stated otherwise; fishing vessels, traditional ships, and recreational crafts with a length of 45 meters or more; ships with bunkers of 5,000 tons or more; and any ship, irrespective of size, carrying dangerous or polluting goods.

International Network for Cooperation and Coordination of Fisheries-Related Monitoring, Control and Surveillance Activities (MCS)

- *Status:* Operational
- *Administration:* Loose network of governmental MCS organizations and others, currently based at NOAA
- *Nature of information:* Serves to improve information collection and exchange among national organizations and institutions responsible for fisheries-related MCS
- *Scope of operation:* Potentially global; to date 40 countries and the European Commission are represented
- *Legal basis:* Voluntary arrangement

INFORMATION ON CONTAINER SECURITY

Central Automated (Cargo/Customs) Information System

- *Status:* Not yet operational; part of the administrative arrangements under the International Convention on Mutual Administrative Assistance in Customs Matters, 2003 (not yet in force)
- *Administration:* WCO Council
- *Nature of information:* Customs-relevant information, including "any other information that may be relevant . . . for ensuring the security of the international trade supply chain" (Art. 30, 27)
- *Scope of operation:* Dependent on the scope of country's participation in the Convention
- *Legal basis:* International Convention on Mutual Administrative Assistance in Customs Matters, 2003, Arts. 27-41

OTHER

There exist other regional maritime information exchange systems that directly bear on maritime security. Mention will be made here of only one—the SOUTHCOM Information Exchange System, based on the DoS Cooperating Nations Information Exchange System,[2] which includes the following:

- The Caribbean Information Sharing Network (CISN) program assists militaries and law enforcement agencies within the Caribbean basin in establishing a community of interest information-sharing network that will enhance bilateral

[2]All the following details are taken directly from <http://68.166.42.251/southcom/ConferencesAndWorkshops/Bahamas23-27Jun,2003/Presentations/C_-_Caribbean_Information_Sharing_Network_(CISN).pdf=>. Accessed on June 25, 2007.

and multilateral cooperation in combating transnational threats and addressing issues of common concern.

- The South American Information Sharing Network (SURNET) provides a community of interest or a regional sensitive-but-unclassified (SBU) protected-information-sharing capability that permits a collaborative approach for addressing transnational threats and other issues of common interest to the South American Military Joint Staff Intelligence Directors.
- REDICA (Central American Network), also known as CENTAM Net, is a community-of-interest initiative to share SBU information in a protected environment among nations in Central America to enhance communication and increase regional cooperation for the purpose of combating common threats and addressing issues of common concern.

G

Acronyms and Abbreviations

AFRICOM	U.S. Africa Command
AIS	Automatic Identification System
AOR	area of responsibility
ASEAN	Association of Southeast Asian Nations
ASP	application service provider
ATA	antiterrorism assistance
BAMS	broad area maritime surveillance
C2	command and control
C2PC	command and control personal computer
CARAT	Cooperation and Afloat Readiness and Training
CBP	Customs and Border Protection
CCG	Commandant of the Coast Guard
CDC	Centers for Disease Control and Prevention
CDVD/IE	Common Distributed Virtual Database/Information Extraction
CENTCOM	U.S. Central Command
CENTRIXS	Combined Enterprise Regional Information Exchange System
CIP	Inter-American Committee on Ports
CISN	Caribbean Information Sharing Network
CMA	Comprehensive Maritime Awareness
CMC	Commandant of the Marine Corps
CNIES	Cooperating Nations Information Exchange System
CNO	Chief of Naval Operations
COCOM	combatant commander

COI	community of interest
COLREG	Convention on the International Regulations for Preventing Collisions at Sea
CONOPS	Concept of Operations
CONUS	continental United States
COP	common operational picture
COTS	commercial off-the-shelf
CSI	Container Security Initiative
CSP	communications service provider
CSR	continuous synopsis record
CST	Caribbean support tender
CSTARS	Center for Southeastern Tropical Advanced Remote Sensing
CTF	Combined Task Force
C–TPAT	Customs–Trade Partnership Against Terrorism
CUI	Controlled Unclassified Information
CZ	contiguous zone
DARPA	Defense Advanced Research Projects Agency
D/B	database
DC	data center
DDOS	distributed denial of service
DEA	Drug Enforcement Administration
DECC	Defense Enterprise Computing Center
DHS	Department of Homeland Security
DMC	disaster monitoring constellation
DOC	Department of Commerce
DOD	Department of Defense
DOE	Department of Energy
DOJ	Department of Justice
DoS	Department of State
DOT	Department of Transportation
DS	data sharing
DSC	digital selective calling
DSCA	Defense Security Cooperation Agency
EEZ	exclusive economic zone
ELINT	electronic intelligence
EPA	Environmental Protection Agency
ESA	European Space Agency
EUCOM	U.S. European Command
EXBS	Export Control and Border Security
FAA	Federal Aviation Administration

FAO	foreign area officer; Food and Agriculture Organization (UN)
FASTC2AP	Fast Connectivity for Coalition Agents Program
FBI	Federal Bureau of Investigation
FIGIS	Fisheries Global Information System
FOC	final operational capability
FSA	Formal Safety Assessment
GENSER	General Service
GISIS	Global Integrated Shipping Information System
GMDSS	Global Maritime Distress and Safety System
GMII	Global Maritime Intelligence Integration
GMP	global maritime partnerships
GMSA	global maritime situation awareness
GOTS	government off-the-shelf
GT	gross ton
GWOT	global war on terror
HARTS	Harbor Craft Transponder System
HHS	Health and Human Services, Department of
HLS	homeland security
HSV	high-speed vessel
HSVAR	High Seas Vessel Authorization Record
ICAO	International Civil Aviation Organization
ICE	Immigration and Customs Enforcement
ILO	International Labor Organization
IMET	international military education and training
IMO	International Maritime Organization
INMARSAT	International Maritime Satellite
IOC	initial operational capability
IP	Internet Protocol
ISPS	International Ship and Port Facility Security
IT	information technology
ITP	International Training Program
JCTD	Joint Concept (Capability) Technology Demonstration
JDL	Joint Directors of Laboratories
JIATF-S	Joint Interagency Task Force-South
JTD	Joint Tactical Demonstration
LE	law enforcement
LEA	law enforcement agencies
LEDET	law enforcement detachment (USCG)

LRIT	Long-Range Identification and Tracking (system)
MAGNet	Maritime Awareness Global Network
MALSINDO	Malaysia, Singapore, and Indonesia
MARAD	Maritime Administration
MDA	maritime domain awareness
MDA DS COI	MDA data-sharing community of interest
MIO	multinational interception operation; maritime interdiction operation
MOS	military occupation specialty
MOTR	Marine Operational Threat Response
MPA	maritime patrol aircraft
MSP	maritime security partnerships
MSPCC	Maritime Security Policy Coordinating Committee
MSSI	Malacca Strait Security Initiative
MSSIS	Maritime Safety and Security Information System
NAIS	Nationwide Automatic Identification System
NATO	North Atlantic Treaty Organization
NAVEUR	U.S. Naval Forces, Europe
NCES	Net-Centric Enterprise Services
NCIS	Naval Criminal Investigative Service
NGO	nongovernmental organization
NMIC	National Maritime Intelligence Center
NOA	notice of arrival
NOAA	National Oceanic and Atmospheric Administration
NORTHCOM	U.S. Northern Command
NRO	National Reconnaissance Office
NSC	National Security Council
NSMS	National Strategy for Maritime Security
NSPD	National Security Presidential Directive
NVOCC	nonvessel operating common carrier
OAE	Operation Active Endeavor
OAS	Organization of American States
ODNI	Office of the Director of National Intelligence
OMP	Office of Manpower and Personnel
ONI	Office of Naval Intelligence
OSD	Office of the Secretary of Defense
ONR	Office of Naval Research
OTH	over the horizon
OTH-B	over-the-horizon-backscatter
OTHR	over-the-horizon radar

P&I	protection and indemnity
PACOM	U.S. Pacific Command
PANDA	Predictive Analysis for Naval Deployment Activities
PCL	printer control language
PEO	program executive office
PFP	Partnership for Peace
PSI	Proliferation Security Initiative
QoS	quality of service
RDT&E	research, development, test, and evaluation
ReCAAP	Regional Cooperation Agreement on Combating Piracy and Armed Robbery against Ships in Asia
REDICA	Central American Information Sharing Network
ReMIX	Regional Maritime Information Exchange
RF	radio frequency
RFMO	Regional Fisheries Management Organization
RIMPAC	Rim of the Pacific
RMAC	Regional Maritime Awareness Capability
RMP	Regional Maritime Partnership
RMSP	Regional Maritime Security Program
ROEs	rules of engagement
ROTHR	relocatable over-the-horizon radar
SA	situational awareness
SAR	search and rescue
SBU/CUI	Sensitive But Unclassified/Controlled Unclassified Information
SCI	sensitive compartmented information
SEI	specific emitter identification
SLOC	sea lane of commerce
SOA	service-oriented architecture
SOLAS	International Convention for the Safety of Life at Sea
SOUTHCOM	U.S. Southern Command
SRS	Ship Reporting System
SSL	Secure Sockets Layer
SSP	sea situation picture
SUA	Suppression of Unlawful Acts (Convention)
SURNET	South American Information Sharing Network
TDMA	time division multiple access
TLS	Transport Layer Security
TRIM	Translingual Instant Messaging
TSC	theater security cooperation

UAV	unmanned aerial vehicle
UN	United Nations
UNCLOS	United Nations Convention on the Law of the Sea
UNITAS	United States–South American Allied Exercise
USAID	U.S. Agency for International Development
USCG	U.S. Coast Guard
USCOM	U.S. Communications Board
USG	U.S. government
USNS	U.S. Naval Ship (civilian-manned)
VHF	very high frequency
VMS	Vessel Monitoring System
VPN	Virtual Private Network
VTC	vessel traffic center
VTS	vessel traffic services; Vessel Tracking System
WCO	World Customs Organization
WMD	weapons of mass destruction
WPNS	Western Pacific Naval Symposium